Student's Solutions Manual

Roxane Barrows
Hocking College

Cheryl Mansky
Hocking College

Introduction to Technical Mathematics
Fifth Edition

Allyn J. Washington
Dutchess Community College

Mario F. Triola
Dutchess Community College

Ellena E. Reda
Dutchess Community College

PEARSON

Addison
Wesley

Boston San Francisco New York
London Toronto Sydney Tokyo Singapore Madrid
Mexico City Munich Paris Cape Town Hong Kong Montreal

Reproduced by Pearson Addison-Wesley from electronic files supplied by the author.

Copyright © 2008 Pearson Education, Inc.
Publishing as Pearson Addison-Wesley, 75 Arlington Street, Boston, MA 02116.

ISBN-13: 978-0-321-37419-6
ISBN-10: 0-321-37419-3

2 3 4 5 6 OPM 10 09 08 07

Table of Contents

Chapter 1: Signed Numbers 1

Chapter 2: Units of Measurement and Approximate Numbers 10

Chapter 3: Introduction to Algebra 14

Chapter 4: Simple Equations and Inequalities 24

Chapter 5: Graphs 38

Chapter 6: Introduction to Geometry 55

Chapter 7: Simultaneous Linear Equations 62

Chapter 8: Factoring 72

Chapter 9: Algebraic Fractions 78

Chapter 10: Exponents, Roots and Radicals 89

Chapter 11: Quadratic Equations 99

Chapter 12: Exponential and Logarithmic Functions 107

Chapter 13: Geometry and Right Triangle Trigonometry 118

Chapter 14: Oblique Triangles and Vectors 127

Chapter 15: Graphs of Trigonometric Functions 143

Chapter 16: Complex Numbers 161

Chapter 17: Introduction to Data Analysis 170

Appendix A-1

Chapter 1
Signed Numbers

1.1 Signed Numbers

1.

3.

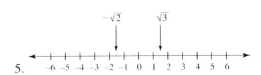

5.

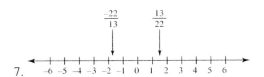

7.

9. 2, farthest to the right

11. -1 ; farthest to right on number line

13. $6 > 2$; 6 is farthest to the right

15. $0 < 4$, 0 is farthest to the right

17. $-3 > -7$

19. $-7 < -5$

21. $\sqrt{5} \approx 2.24 > 2.2$

23. $|6| = 6$ and $|-6| = 6$
    So, $|6| = |-6|$

25. $|6| = 6$ distance from 0, $|-6| = 6$ distance from 0

27. $\left|\dfrac{-6}{7}\right| = \dfrac{6}{7} ; \left|\dfrac{8}{5}\right| = \dfrac{8}{5}$

29. $-\$30$

31. -8 m below

33. answers will vary

35. $-30°C < -5°C$

37. $-2V > -5V$

39. a) -890 ft, -1425ft   b) -1425ft

41. a) 100 V
       $|100| = 100$

    b) $|-200| = 200$
       $|-200|$ V is greater

1.2 Addition and Subtraction of Signed Numbers

1. $2 + 9 = 11$; the sum of two positive numbers is positive

3. $-6 + -9 = -15$; the sum of two negative numbers is negative

5. $-3 + 6 = 3$; unlike signs, subtract and keep the sign of the larger number in absolute value which would be the 6

7. $2 + -10 = 8$; unlike signs, subtract and keep the sign of the larger number in absolute value which would be the -10

9. $12 - 5 = 7$; subtraction sign changes to addition sign, the positive 5 changes to negative 5, then follow rules of addition. $12 = (-5)$. Adding two unlike signs, subtract and keep the sign of the larger one, 12.

11. $7 - 10 = -3$; change to $7 + -10$

13. $-6 - 4 = -10$; change to $-6 + -4$

15. $7 + 9 = 16$

17. $-6 + 7 = 1$

19. $1 + -5 + -2$; add left to right
$-4 + -2 = -6$

21. $2 + -8 - 2$; add left to right
$-6 - 2$; change to addition
$-6 + -2 = -8$

23. $5 - -3 - 7$; add left to right
$5 + 3 - 7$
$8 - 7 = 1$

25. $-7 - -15 + -2$; add left to right
$-7 + 15 + -2$
$8 + -2 = 6$

27. $-9 - -5 + -8 - 5$; add left to right
$-9 + 5 + -8 - 5$
$-4 + -8 - 5$
$-12 - 5 = -17$

29. $3 - -7 - 9 + -3$; add left to right
$10 - 9 + -3$
$1 + -3 = -2$

31. $-6 - 9 - -12 + -4 - -1$; add left to right
$-15 - -12 + 4 - -1$
$-3 + -4 - -1$
$-7 - -1 = -6$

33. a) $-10 \ C + 5 \ C = -5 \ C$

b) $-10 \ C - 5 \ C = -15 \ C$

35. $12 \ A - -5 \ A = 17 \ A$

37. $5000 + 2000 - 3000 + 1000 = 5000$

39. $-8 + 3 = -5$

41. $3 - -8 = 11$

43. $8 - -5 + -1 = 12$

45. $12 - -20 - -15 + -3 = 44$

1.3 Multiplication and Division of Signed Numbers

1. -63; odd number of negatives gives negative product

3. -84; odd number of negatives gives negative product

5. 30; even number of negatives gives positive product

7. 0; zero multiplication rule

9. -56; odd number of negatives

11. 30; even number of negatives

13. -168; odd number of negatives

15. -8; odd number of negatives

17. -3; odd number of negatives

19. -11; odd number of negatives

21. 15, even number of negatives

23. 0; zero can be in the numerator

25. undefined; can't divide by 0

27. $\dfrac{8(-4)}{2} = \dfrac{-32}{2} = -16$

29. $\dfrac{(-5)(15)}{25(-1)} = \dfrac{-75}{-25} = 3$

31. $\dfrac{(-3)(-24)}{-2(-12)} = \dfrac{72}{24} = 3$

33. $\dfrac{10(-6)(-14)}{8(-21)} = \dfrac{840}{-168} = -5$

35. negative; odd number of negatives

37. negative; odd number of negatives

39. positive; even number of negatives

41. zero, multiply by 0

43. 5 s

45. 5 days; $\dfrac{-10 \text{ C}}{-2 \text{ C}}$

47. $-3 \times -6 = 18$

49. $-7 \times 8 = -56$

51. $-63 \div -3 = 21$

53. $12 \div 0 =$ error, undefined

55. $-108 \div 12 = -9$

57. $(2 \times 3 \times 4) \div (-2 \times -6) = 2$

1.4 Powers and Roots

1. 8^3; there are three 8's

3. 2^4; there are four 2's

5. 3^5; there are five 3's

7. 10^5; there are five 10's

9. 8×8

11. $-3 \times 3 \times 3 \times 3 \times 3 \times 3$

13. $7 \times 7 \times 7 \times 7 \times 7 \times 7 \times 7 \times 7$

15. $5 \times 5 \times 5 \times 5 \times 5 \times 5$

17. $-243; -3 \times 3 \times 3 \times 3 \times 3$

19. $64; 4 \times 4 \times 4$

21. $0.09; 0.03 \times 0.03$

23. $42.875; 3.5 \times 3.5 \times 3.5$

25. 4; since $4^2 = 16$

27. 11; since $11^2 = 121$

29. $-4; \sqrt[3]{(-64)} = \sqrt[3]{-4 \times -4 \times -4}$

31. $2; \sqrt[5]{32} = \sqrt[5]{2 \times 2 \times 2 \times 2 \times 2}$

33. $0.4; \sqrt{0.16} = \sqrt{0.4 \times 0.4}$

35. $0.3; \sqrt{0.09} = \sqrt{0.3 \times 0.3}$

37. 108, (27)(4)

39. $-4500; -125 \times 36$

41. 891; (11)(81)

43. 0.637

45. 70,000 J; 7(10,000)

47. $\dfrac{15}{4}$ or 3.75; $\dfrac{\sqrt{225}}{4}$

49. $10,737,418.24$; $2^{30} \div 100$

51. 484 ft; $16(5.5)^2$

53. 125.340; $(0.586)^2 (365)$

55. 27 in; $\sqrt{(20.4)^2 - (17.5)^2} = \sqrt{722.41}$

1.5 Order of Operations

1. $4 - (-8)(2)$

   $4 - -16$

   20

3. $(7.95 \times 10^4)(8.54 \times 10^{-3})$

   $67.893 \times 10^{4 + -3}$

   67.893×10^1

   $6.7893 \times 10^{1+1}$

   6.7893×10^2

5. $(-1)(7) + (-6)(7)$

   $-7 + -42$

   -49

7. $\dfrac{-22}{2} - (5)(-2)$

   $-11 + 10$

   -1

9. $(-8)(9) - \dfrac{-9}{3}$

   $-72 + 3$

   -69

11. $\dfrac{-18}{2} - \dfrac{24}{-6}$

    $-9 + 4$

    -5

13. $(-7)(15) - (-1)^2$

    $-105 - 1$

    -106

15. $4^2 - 2(3) \div (4 - 2)$

    $16 - 2(3) \div 2$

    $16 - 6 \div 2$

    $16 - 3$

    13

17. $-2(-9) - (-1)^4 + \dfrac{-18}{6}$

    $18 - 1 + -3$

    $17 + -3$

    14

19. $\dfrac{6 + 7 \times 2}{10 - 7}$

    $\dfrac{6 + 14}{10 - 7}$

    $\dfrac{20}{3}$

21. $-20 \div -5 + 36 \div -4$

    $4 + -9$

    -5

23. $120 - (8)(3) + 12$

    $120 - 24 + 12$

    108

25. $\dfrac{6(14 - 20)}{3(5) - 6}$

    $\dfrac{6(-6)}{15 - 6}$

    $\dfrac{-36}{9}$

    -4

27. $26 + 8 \div 2^3 - 3(-3)$

    $26 + 8 \div 8 - 3(-3)$

    $26 + 1 + 9$

    36

29. $-1000 \text{ ft} \times 4 + 500 \text{ ft} \times 6$
$-4000 \text{ ft} + 3000 \text{ ft}$
-1000 ft

31. $\dfrac{-9.75 \times 4^2}{2}$
$\dfrac{-156}{2}$
-78 m

33. $\dfrac{102(18.05 - 15.02)}{15.02 - 8.33}$
$\dfrac{102(3.03)}{6.69}$
$\dfrac{309.06}{6.69}$
46.197

35. $2000 - 120 \times 6 + 300(4)$
$2000 - 720 + 1200$
2480 gal

37. $15 \div (10 - 7) - (-2 \div -1) = 3$

39. $5 \wedge 4 \div (6 - 8) + (9 - 4) \div 3 \wedge 2 = -311.94$

41. $(-2) \wedge 6 - 3 \wedge 6 - \left[(-2) \wedge 5 - 2 \wedge 3\right] \div (-2 \times -2 \times 5) = -663$

1.6 Scientific Notation

1. $4 \times 1{,}000{,}000 = 4{,}000{,}000$

3. $8 \times \dfrac{1}{100} = 0.08$

5. $2.17 \times 1 = 2.17$

7. $3.65 \times \dfrac{1}{1000} = 0.00365$

9. 3.000×10^3; moved the decimal point 3 places to the left

11. 7.6×10^{-2}; moved the decimal point 2 places to the right

13. 7.04×10^{-1}; moved the decimal point 1 places to the right

15. 9.21×10^0

17. 5.3×10^{-5}; moved the decimal point 5 places to the right

19. 2.010×10^9; moved the decimal point 9 places to the left

21. $(6.7 \times 10^3)(2.32 \times 10^4)$
$15.544 \times 10^{3+4}$
15.544×10^7
$1.5544 \times 10^{7+1}$
1.5544×10^8

23. $(1.53 \times 10^{-2})(6.08 \times 10^{-1})$
$9.3024 \times 10^{-2+-1}$
9.3024×10^{-3}

25. $725 - 12.50(40)$

$725 - 500$

225

$\dfrac{225}{18.75} = 12$ hour overtime

27. $\dfrac{1.86 \times 10^{-2}}{6.65 \times 10^{-5}}$

$0.2797 \times 10^{-2-(-5)}$

0.2797×10^{3}

$2.797 \times 10^{3-1}$

2.797×10^{2}

29. $(1.037 \times 10^{5}) + (8.364 \times 10^{3})$

$(1.037 \times 10^{5}) - (0.08364 \times 10^{5})$

1.12×10^{5}

31. $\dfrac{3.85 \times 10^{-5}}{9.03 \times 10^{-7}}$

0.426×10^{2}

4.26×10^{1}

33. $\dfrac{(7.53 \times 10^{-2})(7.39 \times 10^{4})}{8.11 \times 10^{-5}}$

$\dfrac{55.6467 \times 10^{2}}{8.11 \times 10^{-5}}$

6.86×10^{7}

35. 9.1×10^{-28} g

37. $5 \times 10^{4} = 5 \times 10,000 = 50,000$ lb

39. 3.6×10^{8} km^{2}

41. 0.0000000000016 W

43. 7.10×10^{8} yr

45. 6.0×10^{-19} J

47. $f = \dfrac{3.00 \times 10^{10}}{4.95 \times 10^{2}}$

$f = 0.6061 \times 10^{8}$

$f = 6.061 \times 10^{7}$ Hz

49. $P = (3.75 \times 10^{-3})(7.24 \times 10^{-4})$

$P = 27.15 \times 10^{-7}$

$P = 2.715 \times 10^{-6}$

51. $(3.0 \times 10^{8})(2.0 \times 10^{-13})$

6×10^{-5}

53. $(2.5 \times 10^{8})(2.6 \times 10^{1})$

6.5×10^{9} gal; 6,500,000,000 gal

1.7 Problem Solving Strategies

1. $\dfrac{390 \text{ mi}}{65 \text{ mi/hr}} = 6$ hr

3. $\dfrac{20 \text{ breaths}}{\text{min}} \times 60 = 1200$ breaths/h

1200×0.5 L $= 600$ L

5. The mean $= \bar{X} = \dfrac{\sum X}{n}$

$69 = \dfrac{550 + 2X}{10}$

$690 = 550 + 2X$

$140 = 2X$

$70 = X$

Each student scored a 70 on the test

7. The flagpole and rope form a right triangle with the rope being the hypotenuse. Since the hypotenuse cannot equal a leg, the flagpoles have 0 ft between them.

9. There are 280 cat owners. So $416 - 280 = 136$

Review Exercises

1. $4 + -6 = -2$; adding unlike signs

3. $-5 - 8 = -13$

5. -63; odd number of negatives

7. 9; even number of negatives

9. $(-3) - (-9) + 4$
   $-3 + 9 + 4$
   $6 + 4$
   10

11. $9 - 2 - (-12)$
    $7 - (-12)$
    19

13. $(-2)(-6)(8)$
    96

15. -60; odd number of negatives

17. $\dfrac{30}{3} = 10$

19. $\dfrac{-72}{12} = -6$

21. $(-4)^4 = -4 \times -4 \times -4 \times -4 = 256$

23. $\dfrac{225}{45} = 5$

25. $-8 + 3 + 5$
    $-5 + 5$
    0

27. $\dfrac{6}{-6} + -8 + 2$
    $-1 + -8 + 2$
    $-9 + 2$
    -7

29. $7 + 12 - 9$
    $19 - 9$
    10

31. $-\dfrac{6}{-6} + -8 - 3^2$
    $1 + -8 - 9$
    -16

33. $9 - 13$

 -4

35. $\dfrac{6 + 14}{10 - 7}$

 $\dfrac{20}{3}$ or $6.\overline{6}$

37. $4 + -9$

 -5

39. 9.805×10^9

41. 3.5×10^{-4}

43. 7.07×10^{-1}

45. $302{,}000{,}000$

47. 0.00187

49. 0.0000000000091

51. $(3.06 \times 10^4) + (4.07 \times 10^6)$

 $(0.0306 \times 10^6) + (4.07 \times 10^6)$

 4.1006×10^6

53. $(7.3 \times 10^7)(3.5 \times 10^{-11})$

 25.55×10^{-4}

 2.555×10^{-3}

55. $6 \text{ qt} - 3 \text{ qt} + 2 \text{ qt} = 5 \text{ qt}$

57. $4 + 22 = 26 \text{ ft}$

59. $-0.1\% + 0.3\% - 0.4\% = -0.2\%$

61. $\dfrac{990{,}000}{600} = 1650$

63. $-8 - 2(4) + 3(3)$

 $-7 \quad C$

65. $\dfrac{87{,}600{,}000 \text{ mi}}{180{,}000 \text{ mi/s}} = 487 \text{ s}$

Chapter Test

1. $-3 + -5 = -8$

2. $-2 + 7 = 5$

3. $-8 - 3 = -11$

4. 24; even number of negatives

5. $7 - (4 - 2) - 1$

 $7 - 2 - 1$

 4

6. $\left(\dfrac{-1}{2}\right)^3 = \dfrac{-1}{2} \times \dfrac{-1}{2} \times \dfrac{-1}{2} = \dfrac{-1}{8}$

7. 3; even number of negatives

8. $12 - 8(4 + 1)$

 $12 - 8(5)$

 $12 - 40$

 -28

9. $-3 + 2^3 - 7$

    $-3 + 8 - 7$

    -2

10. $\dfrac{(4.55)^2 - \sqrt{27}}{(3.2)^3}$

    $\dfrac{20.7025 - 5.196152423}{32.768}$

    0.47

11. $(12.25 + 342.7405928)^3$

    $(354.9905928)^3$

    $44,735,318.5$

12. $\sqrt{\left(3.4^3 + 100\right)}$

    $\sqrt{(39.304 + 100)}$

    $\sqrt{139.304}$

    11.8

13. 1.7×10^{-12}

14. 5.67×10^6

15. 0.000000035

16. $12,400$

17. $(4.5 \times 10^{-4}) + (5.9 \times 10^{-2})$

    $(0.045 \times 10^{-2}) + (5.9 \times 10^{-2})$

    5.945×10^{-2}

18. $(4.92 \times 10^4) - (3.21 \times 10^3)$

    $(4.92 \times 10^4) - (0.321 \times 10^4)$

    4.599×10^4

19. $(2.46 \times 10^{-6})(5.9 \times 10^{-3})$

    14.514×10^{-9}

    1.4514×10^{-8}

20. $\dfrac{2.29 \times 10^7}{7.328 \times 10^{-2}}$

    0.3125×10^9

    3.125×10^8

21. $-5 + 7 + 9 - 5 + 15 + 4 - 6 + 20 - 1 + 5 = 43$ yd gain

22. $\dfrac{4.1601 \times 10^{20}}{6.2415 \times 10^{18}}$

    0.666522×10^2

    66.65 coulombs

23. $725 - 12.50(40)$

    $725 - 500$

    225

    $\dfrac{225}{18.75} = 12$ hour overtime

Chapter 2
Units of Measurement
And
Approximate Numbers

2.1 Working with Units of Measure

1. $8 \cancel{h} \times \dfrac{60 \, m\cancel{in}}{1 \, \cancel{h}} \times \dfrac{60 \, s}{1 \, m\cancel{in}} = 28,800 \, s$

3. $26.2 \, m\cancel{i} \times \dfrac{5280 \, f\cancel{t}}{1 \, m\cancel{i}} \times \dfrac{1 \, m\cancel{in}}{515 \, f\cancel{t}} \times \dfrac{1 \, h}{60 \, min} = 4.5 \, h$

5. $85 \, \cancel{lb} \times \dfrac{16 \, oz}{1 \, \cancel{lb}} = 1,360 \, oz$

7. $\dfrac{88 \, f\cancel{t}}{1 \, \cancel{s}} \times \dfrac{1 \, mi}{5280 \, f\cancel{t}} \times \dfrac{3600 \, \cancel{s}}{1 \, h} = 60 \, mi/h$

9. $\dfrac{65 \, m\cancel{i}}{1 \, \cancel{h}} \times \dfrac{5280 \, ft}{1 \, m\cancel{i}} \times \dfrac{1 \, \cancel{h}}{3600 \, s} = 95.33 \, ft/s$

11. $\dfrac{\$45,000}{1 \, y\cancel{r}} \times \dfrac{1 \, y\cancel{r}}{52 \, w\cancel{k}} \times \dfrac{1 \, w\cancel{k}}{40 \, h} = \$21.63/h$

13. $26.2 \, m\cancel{i} \times \dfrac{5280 \, f\cancel{t}}{1 \, m\cancel{i}} \times \dfrac{1 \, m\cancel{in}}{515 \, f\cancel{t}} \times \dfrac{1 \, h}{60 \, min} = 4.5 \, h$

15. $\dfrac{37,000 \, f\cancel{t}}{1 \, \cancel{s}} \times \dfrac{1 \, mi}{5280 \, f\cancel{t}} \times \dfrac{3600 \, \cancel{s}}{1 \, h} = 25,227.27 \, mi/h$

2.2 Units of Measurement: The Metric System

1. mA; 1mA = 0.001 A

3. kV; 1 kV = 1000 V

5. kW; 1 kW = 1000 W

7. ML; 1 ML = 1,000,000 L

9. megavolt; 1 MV = 1,000,000 V

11. microsecond; $\mu s = 0.000001 \, s$ $\mu s = 0.000001 \, s$

13. centivolt; 1 cV = 0.01 V

a) $\dfrac{765 \, km}{1 \, \cancel{h}} \times \dfrac{1 \, \cancel{h}}{5650 \, L} = 0.135 \, km/L$

b) $\dfrac{5650 \, L}{1 \, \cancel{h}} \times \dfrac{1 \, \cancel{h}}{765 \, k\cancel{m}} \times \dfrac{2850 \, k\cancel{m}}{230 \, passengers} = 91.5 \, L/passenger$

15. nanoampere; 1 nA = 0.000000001 A

17. a) m^2

   b) ft^2

19. a) m/s

   b) ft/s

21. 4 s

23. 8 m

25. $0.04 \, \cancel{s} \times \dfrac{100 \text{ cs}}{1 \, \cancel{s}} = 4 \text{ cs}$

27. $0.000003 \, \cancel{s} \times \dfrac{100,000,000 \, \mu s}{1 \, \cancel{s}} = 3 \, \mu s$

29. $0.08 \, \cancel{g} \times \dfrac{100 \text{ cg}}{1 \, \cancel{g}} = 8 \text{ cg}$

31. $4000 \, \cancel{g} \times \dfrac{1 \text{ kg}}{1000 \, \cancel{g}} = 4 \text{ kg}$

33. mm^2

35. mi/gal

37. $\dfrac{1}{s}$

39. $\dfrac{\text{lb}}{\text{ft}^2}$

41. $\dfrac{\text{kg} \times \text{m}}{\text{s}^2}$

43. As

2.3 Reduction and Conversion of Units

1. $1 \, \cancel{\text{cm}} \times \dfrac{10 \text{ mm}}{1 \, \cancel{\text{cm}}} = 10 \text{ mm}$

3. $1 \, \cancel{\text{mi}} \times \dfrac{5280 \, \cancel{\text{ft}}}{1 \, \cancel{\text{mi}}} \times \dfrac{12 \text{ in}}{1 \, \cancel{\text{ft}}} = 63,360 \text{ in}$

5. $2 \, \cancel{\text{ft}}^2 \times \dfrac{144 \text{ in}^2}{1 \, \cancel{\text{ft}}^2} = 288 \text{ in}^2$

7. $12 \, \cancel{\text{in}} \times \dfrac{2.54 \text{ cm}}{1 \, \cancel{\text{in}}} = 30.48 \text{ cm}$

9. $55 \, \cancel{\text{gal}} \times \dfrac{4 \text{ qt}}{1 \, \cancel{\text{gal}}} = 220 \text{ qt}$

11. $25 \, \cancel{\text{qt}} \times \dfrac{1 \text{ L}}{1.057 \, \cancel{\text{qt}}} = 23.7 \text{ L}$

13. $5.2 \, \cancel{\text{m}}^2 \times \dfrac{10,000 \text{ cm}^2}{1 \, \cancel{\text{m}}^2} = 52,000 \text{ cm}^2$

15. $256.3 \, \cancel{\text{L}} \times \dfrac{1 \text{ ft}^3}{28.32 \, \cancel{\text{L}}} = 9.05 \text{ ft}^3$

17. $\dfrac{80 \, \cancel{\text{km}}}{1 \text{ h}} \times \dfrac{1 \text{ mi}}{1.609 \, \cancel{\text{km}}} = 50 \text{ mi/h}$

19. $4 \, \cancel{\text{qt}} \times \dfrac{1 \text{ L}}{1.057 \, \cancel{\text{qt}}} = 3.8 \text{ L}$

21. $\dfrac{770 \, \cancel{\text{mi}}}{1 \, \cancel{\text{h}}} \times \dfrac{5280 \text{ ft}}{1 \, \cancel{\text{mi}}} \times \dfrac{1 \, \cancel{\text{h}}}{3600 \text{ s}} = 1129 \text{ ft/s}$

23. $\dfrac{5.52 \, \cancel{\text{g}}}{1 \, \cancel{\text{cm}}^3} \times \dfrac{1 \text{ kg}}{1000 \, \cancel{\text{g}}} \times \dfrac{1,000,000 \, \cancel{\text{cm}}^3}{1 \text{ m}^3} = 5520 \text{ kg/m}^3$

25. $400 \, \cancel{\text{m}} \times \dfrac{1 \, \cancel{\text{km}}}{1000 \, \cancel{\text{m}}} \times \dfrac{1 \text{ mi}}{1.609 \, \cancel{\text{km}}} = 0.25 \text{ mi}$

27. $100 \, \cancel{\text{yd}} \times \dfrac{36 \, \cancel{\text{in}}}{1 \, \cancel{\text{yd}}} \times \dfrac{1 \text{ m}}{39.37 \, \cancel{\text{in}}} = 91.4 \text{ m}$

29. $\dfrac{3.47 \, \cancel{\text{oz}}}{1 \, \cancel{\text{gal}}} \times \dfrac{1 \text{ lb}}{16 \, \cancel{\text{oz}}} \times \dfrac{1 \, \cancel{\text{gal}}}{4 \, \cancel{\text{qt}}} \times \dfrac{1.057 \, \cancel{\text{qt}}}{1 \, \cancel{\text{L}}} \times \dfrac{28.32 \, \cancel{\text{L}}}{1 \text{ ft}^3} = 1.6 \text{ lb/ft}^3$

2.4 Approximate Numbers and Significant Digits

1. exact

3. approximate

5. approximate

7. 1 cm and 1 mm are approximate; $2.80 is exact

9. 563 has 3 significant digits; 4029 has 4 significant digits

11. 3799 has 4 significant; 2001 has 4 significant digits

13. 5.80 has 3 significant; 5.08 has 3 significant

15. 10,060 has 4 significant; 403,020 has 5 significant

17. a) 3.764; b) 3.764

19. a) 0.01; b) 30.8

21. a) same precision; b) 78.0

23. a) 0.004; b) same accuracy

25. a) 5.71; b) 5.7

27. a) 6.93; b) 6.9

29. a) $41\overline{0}0$; b) 4100

31. a) 46,800; b) 47,000

33. a) 501; b) $5\overline{0}0$

35. a) 0.215; b) 0.22

37. 128.3 ± 0.05; 128.25 ft, 128.35 ft

39. 82 ± 0.5; 81.5 L. 82.5 L

41. $164.0 \text{ m}\cancel{L} \times \dfrac{1 \cancel{L}}{1000 \text{ m}\cancel{L}} \times \dfrac{1.057 \text{ qt}}{1 \cancel{L}} = 0.1733 \text{ qt}$

43. $100 \text{ y}\cancel{d} \times \dfrac{36 \text{ i}\cancel{n}}{1 \text{ y}\cancel{d}} \times \dfrac{2.54 \text{ c}\cancel{m}}{1 \text{ i}\cancel{n}} \times \dfrac{1 \text{ m}}{100 \text{ c}\cancel{m}} = 91.4 \text{ m}$

45. 150.4 + 95.66 + 81 = 327.06; round to least precise gives 327 ft

47. $0.0225 \times 1.657 = 0.0372825$; rounded to least accurate gives 0.0373 W

Review Exercises

1. 4; 4

3. 3; 4

5. a) 7.32 more precise; b) same

7. a) same; b) 207.31

9. a) 98.5; b) 98

11. a) 60,500; b) 61,000

13. a) 673; b) 670

15. a) 0.700; b) 0.70

17. 1 μg = 0.000001 g

19. 1 ks = 1000 s

21. $385 \text{ m}\cancel{m}^3 \times \dfrac{1 \text{ cm}^3}{10^3 \text{ m}\cancel{m}^3} = 0.385 \text{ cm}^3$

23. $5.2 \text{ i}\cancel{n} \times \dfrac{2.54 \text{ cm}}{1 \text{ i}\cancel{n}} = 13 \text{ cm}$

25. $4.452 \text{ gal} \times \dfrac{4 \text{ qt}}{1 \text{ gal}} \times \dfrac{2 \text{ pt}}{1 \text{ qt}} = 35.62 \text{ pt}$ 27. $27 \text{ ft}^3 \times \dfrac{28.32 \text{ L}}{1 \text{ ft}^3} = 760 \text{ L}$

29. $0.43 \text{ ft}^3 \times \dfrac{1728 \text{ in}^3}{1.609 \text{ ft}^3} = 740 \text{ in}^3$

31. $\dfrac{2.45 \text{ mi}}{1 \text{ h}} \times \dfrac{1.609 \text{ km}}{1 \text{ mi}} \times \dfrac{1000 \text{ m}}{1 \text{ km}} \times \dfrac{1 \text{ h}}{3600 \text{ s}} = 1.10 \text{ m/s}$

33. 0.1855 in, 0.1865   35. 21.84 g

37. $0.000012 \times 38 = 0.021 \text{ m}$

39. $\dfrac{5.6 \text{ lb}}{1 \text{ gal}} \times \dfrac{1 \text{ kg}}{2.205 \text{ lb}} \times \dfrac{1 \text{ gal}}{4 \text{ qt}} \times \dfrac{1.057 \text{ qt}}{1 \text{ L}} = 0.67 \text{ kg/L}$

41. $287 \text{ mil}^2 \times \dfrac{(0.001)^2 \text{ in}^2}{1 \text{ mil}^2} = 0.000287 \text{ in}^2$

Chapter Test

1. a) 3; b) 3; c) 3   2. a) 4.36; b) 0.00616; c) 105

3. a) kilovolts; 1 kV = 1000 V; b) milliampere; 1 mA = 0.001 A; c) millimeter; 1 mm = 0.001 m

4. a) $1 \text{ km} \times \dfrac{1000 \text{ m}}{1 \text{ km}} \times \dfrac{100 \text{ cm}}{1 \text{ m}} = 100{,}000 \text{ cm}$

   b) $5.60 \text{ kg} \times \dfrac{2.205 \text{ lb}}{1 \text{ kg}} = 12.3 \text{ lb}$

5. $56 \text{ L} \times \dfrac{1.057 \text{ qt}}{1 \text{ L}} \times \dfrac{1 \text{ gal}}{4 \text{ qt}} = 15 \text{ gal}$ 6. $\dfrac{8.5 \text{ gal}}{1 \text{ min}} \times \dfrac{4 \text{ qt}}{1 \text{ gal}} \times \dfrac{1 \text{ L}}{1.057 \text{ qt}} \times \dfrac{1 \text{ min}}{60 \text{ s}} = 0.54 \text{ L/s}$

7. $\dfrac{3.47 \text{ oz}}{1 \text{ gal}} \times \dfrac{1 \text{ lb}}{16 \text{ oz}} \times \dfrac{1 \text{ gal}}{4 \text{ qt}} \times \dfrac{1.057 \text{ qt}}{1 \text{ L}} \times \dfrac{28.32 \text{ L}}{1 \text{ ft}^3} = 1.62 \text{ lb/ft}^3$

8. $1414 \text{ ft} \times \dfrac{12 \text{ in}}{1 \text{ ft}} \times \dfrac{2.54 \text{ cm}}{1 \text{ in}} \times \dfrac{1 \text{ m}}{100 \text{ cm}} = 431.0 \text{ m}$

   431.0 m is the height of the Empire State Building in meters. It is taller.

9. F = 1.8C + 32
   F = 1.8(100) + 32
   F = 180 + 32
   F = 212

10. a) $\dfrac{765 \text{ km}}{1 \text{ h}} \times \dfrac{1 \text{ h}}{5650 \text{ L}} = 0.135 \text{ km/L}$

    b) $\dfrac{5650 \text{ L}}{1 \text{ h}} \times \dfrac{1 \text{ h}}{765 \text{ km}} \times \dfrac{2850 \text{ km}}{230 \text{ passengers}} = 91.5 \text{ L/passenger}$

Chapter 3
Introduction to Algebra

3.1 Working with Formulas

1. b, c

3. 7, p, q, r

5. i, i, R

7. a, b, c, c, c

9. 6

11. 2π

13. $4\pi e$

15. mw^2

17. $x = 4y$

19. $m = 10v$

21. $V = 7.48lwd$

23. $d = \dfrac{1}{2}gt^2$

25. $N = 5280x$

27. $N = 9xy$

29. $N = 24n$

31. $C = 4sc$

33. $\dfrac{6000 \text{ mi}}{1 \cancel{h}} \bullet \dfrac{1 \cancel{h}}{60 \text{ m}\cancel{in}} \bullet \dfrac{15 \text{ m}\cancel{in}}{1} = 1500 \text{ mi}$

35. $V = IR$
    $V = 0.075 \bullet 20$
    $V = 1.5 \text{ V}$

37. $V = 4 \text{ ft} \bullet 8 \text{ ft} \bullet 3 \text{ ft}$
    $V = 96 \text{ ft}^3$
    $V = 96 \text{ f}\cancel{t}^3 \bullet \dfrac{28.32 \text{ L}}{1 \text{ f}\cancel{t}^3} \bullet \dfrac{1.057 \text{ q}\cancel{t}}{1 \cancel{L}} \bullet \dfrac{1 \text{ gal}}{4 \text{ q}\cancel{t}} = 720 \text{ gal}$

39. $d = \dfrac{1}{2}gt^2$
    $d = \dfrac{1}{2} \bullet 32.2(2.25)^2$
    $d = 81.51 \text{ ft}$

41. $N = 24n$
    $N = 24 \bullet 144$
    $N = 3456 \text{ bolts}$

43. $P = 4(12)(1.35)$
    $P = \$64.80$

3.2 Basic Algebraic Expressions

1. x^2, $4xy$, $-7x$

3. 12, $-5xy$, $7x$, $\dfrac{-x}{8}$

5. 3x and 2x are like terms

7. x and 5x are like terms

9. $-8mn$ and $-mn$ are like terms

11. $6(x-y)$ and $-3(x-y)$ are like terms

13. $6a(a-x)$

15. $x^2(a-x)(a+x)$

17. $\dfrac{2}{5a}$

19. $\dfrac{6}{a-b}$

21. $K = 55 + 273 = 328$

23. $t = \sqrt{\dfrac{64}{16}} = \sqrt{4} = 2$

25. $L = 50 - x$

27. $A = 2x^2 + 4xl$

29. $m = \dfrac{1}{2}(s+1)$

31. $E = \dfrac{I-P}{I}$

33. $L = 50 - 27.4 = 22.6$

35. $A = 2(2.91)^2 + 4(5.23)(2.91) = 77.81 \text{ ft}^2$

37. $V = \dfrac{4000}{3+1} = \$1,000$

39. $E = \dfrac{21500 - 7600}{21500} = 0.6465$

41. $A = 2(1.238)^2 + 4(1.348)(1.238) = 9.741 \text{ in}^2$

3.3 Addition and Subtraction of Algebraic Expressions

1. $6x + y$

3. $7a - 4b^2$

5. $5s + 3t$

7. $2a + 3$

9. $3 - 4x - 7$
   $-4x - 4$

11. $4s - 9 + 8 - 2s$
    $2s - 1$

13. $t - 7 + 3y - 2 + y - t$
    $4y - 9$

15. $4 + [6x - 3 + 4x]$
    $4 + 10x - 3$
    $10x + 1$

17. $3s - [2 - 4 + s]$
    $3s - 2 + 4 - s$
    $2s + 2$

19. $-7 + x - [2x + 3 - 7x + 2]$
    $-7 + x - 2x - 3 + 7x - 2$
    $6x - 12$

21. $t - 5x - \{6p^2 - x - 9 - 6t - p^2 + x\}$
    $t - 5x - 6p^2 + x + 9 + 6t + p^2 - x$
    $7t - 5x - 5p^2 + 9$

23. $\left[\left(a^2+x+1\right)+\left(4a^2-5x+2\right)\right]-(3a^2-7x+2)$

$\qquad a^2+x+1+4a^2-5x+2-3a^2+7x-2$

$\qquad 2a^2+3x+1$

25. $\left\{[3x-t+4t]+[2x+8-5t]\right\}-\left\{(4x+2-3t)+(x-1-t)\right\}$

$\qquad \left\{5x-2t+8\right\}-\left\{5x+1-4t\right\}$

$\qquad 5x-2t+8-5x-1+4t$

$\qquad 2t+7$

27. $(x-a)+(3x+6a)$

$\qquad x-a+3x+6a$

$\qquad 4x+5a$

29. $(5x+3)+(x-7)+(5x+4)$

$\qquad 8x$

31. $(6x+20)-(4x-8)$

$\qquad 6x+20-4x+8$

$\qquad 2x+28$

33. $(5R-20)-(3R-50)$

$\qquad 5R-20-3R+50$

$\qquad 2R+30$

35. $4M-(2M+120)+80$

$\qquad 4M-2M-120+80$

$\qquad 2M-40$

37. $\left(-16t^2+160\right)-\left(-16t^2+80\right)$

$\qquad -16t^2+160+16t^2-80$

$\qquad 80\ \text{ft}$

39. $2(8-x)+2(20-3x)$

$\qquad 16-2x+40-6x$

$\qquad -8x+56$

3.4 Multiplication of Algebraic Expressions

1. $x^{3+7}=x^{10}$

3. $y^{5+2+1}=y^8$

5. $t^{3+5+2}=t^{10}$

7. $n^{2\bullet 7}=n^{14}$

9. $(-a)^2 x^{2\bullet 2}b^2=a^2x^4b^2$

11. $(-a)^5 t^{2\bullet 5}=-a^5t^{10}$

13. $-84r^2s^2t^3$

15. $(-2)^3 s^3 t^{3\bullet 3}x^3=-8s^3t^9x^3$

17. $2a^2+6ax$

19. $6a^2x-3a^4$

21. $-2s^2tx+2st^3y$

23. $-3x^3y^2+9ax^2y^7$

25. $(x-3)(x-1)$

$x^2 - x - 3x + 3$

$x^2 - 4x + 3$

27. $(s-2)(s+3)$

$s^2 + 3s - 2s - 6$

$s^2 + s - 6$

29. $(x+1)(2x-1)$

$2x^2 - x + 2x - 1$

$2x^2 + x - 1$

31. $(5v+1)(2v+3)$

$10v^2 + 15v + 2v + 3$

$10v^2 + 17v + 3$

33. $(a-x)(a-2x)$

$a^2 - 2ax - ax + 2x^2$

$a^2 - 3ax + 2x^2$

35. $(2a-c)(3a+2c)$

$6a^2 + 4ac - 3ac - 2c^2$

$6a^2 + ac - 2c^2$

37. $2x^2 + 18x - 5xt - 45t$

39. $4a^2 + 18apy - 18apy - 81p^2y^2$

$4a^2 - 81p^2y^2$

41. $2a^3 - 6a^2 - 10a + a^2 - 3a - 5$

$23a^3 - 5a^2 - 13a - 5$

43. $a^2 + 2axy - 3ax - ax - 2x^2y + 3x^2$

$a^2 + 2axy - 4ax - 2x^2y + 3x^2$

45. $(x-2)^2 = (x-2)(x-2)$

$x^2 - 2x - 2x + 4$

$x^2 - 4x + 4$

47. $(x+2y)(x+2y)$

$x^2 + 2xy + 2xy + 4y^2$

$x^2 + 4xy + 4y^2$

49. $\left(x^2 - 2x + 3x - 6\right)(x-4)$

$\left(x^2 + x - 6\right)(x-4)$

$x^3 + x^2 - 6x - 4x^2 - 4x + 24$

$x^3 - 3x^2 - 10x + 24$

51. $(x+1)(x+1)(x+1)$

$\left(x^2 + x + x + 1\right)(x+1)$

$\left(x^2 + 2x + 1\right)(x+1)$

$x^3 + 2x^2 + x + x^2 + 2x + 1$

$x^3 + 3x^2 + 3x + 1$

53. a) false; $(2x)^3 = 2^3x^3 = 8x^3$

b) false; $\left(x^2y\right)^3 = x^6y^3$

c) true

55. a) true   b) true   c) false; $x^2y^3 \ne (xy)^6$

57. a) false; $(a+b)^2 = (a+b)(a+b) = a^2 + 2ab + b^2$

b) false; $(x-3)^2 = (x-3)(x-3) = x^2 - 6x + 9$

c) true

59. a) false; $(t+1)(t+1) = t^2 + 2t + 1$

   b) true   c) true

61. $(x+3)(x-2)$

   $x^2 - 2x + 3x - 6$

   $x^2 + x - 6$ ft

63. $200(x+3)(x+5)$

   $200(x^2 + 5x + 3x + 15)$

   $200(x^2 + 8x + 15)$

   $200x^2 + 1600x + 3000$

65. $(2x+y)(2x-y) = 4x^2 - y^2$

   $4x^2 + 2xy - 2xy - y^2 = 4x^2 - y^2$

   $4x^2 - y^2 = 4x^2 - y^2$

   $[2 \bullet 5 + 3][2 \bullet 5 - 3] = 4(5)^2 - 3^2$

   $[10+3][10-3] = 91$

   $91 = 91$

67. $\dfrac{1}{f} = (n-1)(r_2 - r_1)$

   $\dfrac{1}{f} = nr_2 - nr_1 - r_2 + r_1$

3.5 Division of Algebraic Expressions

1. $x^{7-4} = x^3$

3. $a^{5-4} = a$

5. $4n^{4-1} = 4n^3$

7. $\dfrac{-x^4 y^3}{x^2 y} = \dfrac{-x^2 y^2}{1} = -x^2 y^2$

9. $\dfrac{-5x^2 yr}{20xyr^3} = \dfrac{x^{2-1} y^{1-1}}{-4r^{3-1}} = \dfrac{x}{-4r^2}$

11. $\dfrac{9abc^4 ds}{15bc^4 d^4} = \dfrac{3ab^{1-1} c^{4-4} s}{5d^{4-1}} = \dfrac{3as}{5d^3}$

13. $\dfrac{6ab + 5a}{a} = \dfrac{6ab}{a} + \dfrac{5a}{a} = 6b + 5$

15. $\dfrac{9m^2}{3m} - \dfrac{3m}{3m} = 3m - 1$

17. $\dfrac{a^3 x^4}{-ax^2} - \dfrac{a^2 x^3}{-ax^2} = -a^2 x^2 + ax$

19. $\dfrac{3xy^4}{3xy^2} - \dfrac{6x^2 y^5}{3xy^2} = y^2 - 2xy^3$

21. $\dfrac{a^2 b^2 c^3}{a^2 bc^2} - \dfrac{a^3 b^4 c^6}{a^2 bc^2} - \dfrac{2a^3 bc^2}{a^2 bc^2} = bc - ab^3 c^4 - 2a$

23. $\dfrac{a^2 b^3}{-ab} - \dfrac{2a^3 b^4}{-ab} - \dfrac{ab}{-ab} - \dfrac{ab^2}{-ab} = -ab^2 + 2a^2 b^3 + 1 + b$

25.
$$
x+1 \overline{\smash{\big)}\ x^2 - 2x - 3} \quad \underset{\text{over}}{x-3}
$$

$$
\begin{array}{r}
x - 3 \\
x+1 \overline{\smash{\big)}\ x^2 - 2x - 3} \\
\underline{x^2 + x} \\
0 - 3x - 3 \\
\underline{3x - 3} \\
0
\end{array}
$$

27.
$$
\begin{array}{r}
2x + 1 \\
x-3 \overline{\smash{\big)}\ 2x^2 - 5x - 3} \\
\underline{2x^2 - 6x} \\
0 \ + x - 3 \\
\underline{x - 3} \\
0
\end{array}
$$

29.
$$
\begin{array}{r}
4x - 3 + \dfrac{2}{2x+3} \\
2x+3 \overline{\smash{\big)}\ 8x^2 + 6x - 7} \\
\underline{8x^2 + 12x} \\
0 \ - 6x - 7 \\
\underline{-6x - 9} \\
2
\end{array}
$$

31.
$$
\begin{array}{r}
2x^2 - 3x - 4 \\
4x-3 \overline{\smash{\big)}\ 8x^3 - 18x^2 - 7x + 12} \\
\underline{8x^3 - 6x^2} \\
0 \ - 12x^2 - 7x \\
\underline{-12x^2 + 9x} \\
0 \ - 16x + 12 \\
\underline{-16x + 12} \\
0
\end{array}
$$

33.
$$
\begin{array}{r}
2x^2 - x + 3 \\
x-2 \overline{\smash{\big)}\ 2x^3 - 5x^2 + 5x - 6} \\
\underline{2x^3 - 4x^2} \\
0 \ - x^2 + 5x \\
\underline{-x^2 + 2x} \\
0 + 3x - 6 \\
\underline{3x - 6} \\
0
\end{array}
$$

35.
$$
\begin{array}{r}
x^3 + x^2 + x + 1 \\
x-1 \overline{\smash{\big)}\ x^4 + 0x^3 + 0x^2 + 0x - 1} \\
\underline{x^4 - x^3} \\
0 + x^3 + 0x^2 \\
\underline{x^3 - x^2} \\
0 + x^2 + 0x \\
\underline{x^2 - x} \\
0 + x - 1 \\
\underline{x - 1}
\end{array}
$$

37.
$$x+1 \overline{\smash{)}5x^2 + 0x - 5} \quad \begin{array}{c} 5x - 5 \end{array}$$
$$\underline{5x^2 + 5x}$$
$$0 \quad -5x - 5$$
$$\underline{-5x - 5}$$
$$0$$

39.
$$3x - 1 \overline{\smash{)}6x^4 - 5x^3 + 7x^2 + x - 3} \quad \begin{array}{c} 2x^3 - x^2 + 2x + 1 - \dfrac{2}{3x - 1} \end{array}$$
$$\underline{6x^4 - 2x^3}$$
$$0 \quad -3x^3 + 7x^2$$
$$\underline{-3x^3 + x^2}$$
$$0 \quad +6x^2 + x$$
$$\underline{6x^2 - 2x}$$
$$\text{keep dividing}$$

41.
$$a + 4b \overline{\smash{)}a^2 + ab - 12b^2} \quad \begin{array}{c} a - 3b \end{array}$$
$$\underline{a^2 + 4ab}$$
$$0 - 3ab - 12b^2$$
$$\underline{-3ab - 12b^2}$$
$$0$$

43.
$$x - y \overline{\smash{)}x^3 - 5x^2y + 2xy^2 + 2y^3} \quad \begin{array}{c} x^2 - 4xy - 2y^2 \end{array}$$
$$\underline{x^3 - x^2y}$$
$$0 \quad -4x^2y + 2xy^2$$
$$\underline{-4x^2y + 4xy^2}$$
$$0 \quad -2xy^2 + 2y^3$$
$$\underline{-2xy^2 + 2y^3}$$
$$0$$

45. a) false; $x^8 \div x^2 = x^6$ b) false; $\dfrac{r}{r^3} = \dfrac{1}{r^2}$

c) false; $\dfrac{x^6}{x^3} = x^3$

47. a) false; $\dfrac{x^2 + y^2}{x} = \dfrac{x^2}{x} + \dfrac{y^2}{x} = x + \dfrac{y^2}{x}$ b) true   c) true

49. a) true   b) true   c) false; $\dfrac{6x^2 - 8}{x^2} = \dfrac{6x^2}{x^2} - \dfrac{8}{x^2} = 6 - \dfrac{8}{x^2}$

51.
$$r + 2 \overline{\smash{)}6r^2 + 8r + 10} \quad \begin{array}{c} 6r - 4 + \dfrac{18}{r + 2} \end{array}$$
$$\underline{6r^2 + 12r}$$
$$0 \quad -4r + 10$$
$$\underline{-4r - 8}$$
$$18$$

53.
$$x + 2 \overline{\smash{)}3x^2 - 8x - 28} \quad \begin{array}{c} 3x - 14 \end{array}$$
$$\underline{3x^2 + 6x}$$
$$0 \quad -14x - 28$$
$$\underline{-14x - 28}$$
$$0$$

Review Exercises

1. $-7a$ and $5a$

3. $5(a-b)$ and $-7(a-b)$

5. $7a - 4b$

7. $12ax - 19bx$

9. $3x + 3y - 2y$
   $3x + y$

11. $a - x + 2a$
    $3a - x$

13. $2x - 10y - 3y + 7x$
    $9x - 13y$

15. $-2 + n + 4 - 6 + n$
    $-4 + 2n$

17. $2x - 3x + 9y - y + x$
    $8y$

19. $-6a^3b^6$

21. $\left(-8x^3y^6z^3\right)\left(-7xy^3z^5\right)$
    $56x^4y^9z^8$

23. $3^3a^3b^6 = 27a^3b^6$

25. $2^4x^8y^4z^{12} = 16x^8y^4z^{12}$

27. $-4ax^3$

29. $\dfrac{5x}{y^4z^3}$

31. $5 - [x - 3 + 4x]$
    $5 - [5x - 3]$
    $5 - 5x + 3$
    $8 - 5x$

33. $2x + [2x - a - a + x]$
    $2x + 3x - 2a$
    $5x - 2a$

35. $2 - \{2 - [3x - 7 + x]\}$
    $2 - \{2 - [4x - 7]\}$
    $2 - \{2 - 4x + 7\}$
    $2 - \{9 - 4x\}$
    $2 - 9 + 4x$
    $-7 + 4x$

37. $-\{2b - [b - 4 + 5b] + 6 - b\}$
    $-\{2b - b + 4 - 5b + 6 - b\}$
    $-\{-5b + 10\}$
    $5b - 10$

39. $2x^5 - 6x^3$

41. $-2a^3x + 2a^2t$

43. $2x^2 + 7x - 6x - 21$
    $2x^2 + x - 21$

45. $6a^2 + 4ab - 15ab - 10b^2$
    $6a^2 - 11ab - 10b^2$

47. $x^3 - x^2 + x + x^2 - x + 1$
    $x^3 + 1$

49. $(4-2x)(x^2-x-4)$

$4x^2-4x-16-2x^3+2x^2+8x$

$6x^2-2x^3+4x-16$

51. $\dfrac{2x^3y^5}{-x^3y^2}-\dfrac{3x^6y^2}{-x^3y^2}$

$-2y^3+3x^3y^2$

53. $\dfrac{h^2j^4}{hj^4}-\dfrac{3hj^6}{hj^4}-\dfrac{6h^4j^7}{hj^4}$

$h-3j^2-6h^3j^3$

55.
$$
\begin{array}{r}
x+4 \\
x-3\overline{\smash{\big)}\,x^2+x-12} \\
\underline{x^2-3x} \\
0\ +\ 4x-12 \\
\underline{4x-12} \\
0
\end{array}
$$

57.
$$
\begin{array}{r}
x^2-x+1-\dfrac{11}{2x+3} \\
2x+3\overline{\smash{\big)}\,2x^3+x^2-x-8} \\
\underline{2x^3+3x^2} \\
0\ \ -2x^2-x \\
\underline{-2x^2-3x} \\
0\ \ \ \ +2x-8 \\
\underline{2x+3} \\
-11
\end{array}
$$

59.
$$
\begin{array}{r}
x^2-x+1 \\
x+1\overline{\smash{\big)}\,x^3+0x^2+0x+1} \\
\underline{x^3+\ x^2} \\
0\ \ -x^2+0x \\
\underline{-x^2-x} \\
0\ \ +\ x+1 \\
\underline{x+1} \\
0
\end{array}
$$

61.
$$
\begin{array}{r}
2x^2-x\ -4 \\
x^2+2\overline{\smash{\big)}\,2x^4-x^3+0x^2-2x-8} \\
\underline{2x^4+4x^2} \\
0\ -x^3-4x^2-2x \\
\underline{-x^3-2x} \\
0\ -4x^2\ +8 \\
\underline{-4x^2\ +8} \\
0
\end{array}
$$

63.
$$
\begin{array}{r}
x+2y \\
3x-y\overline{\smash{\big)}\,3x^2+5xy-2y^2} \\
\underline{3x^2-xy} \\
0\ +6xy-2y^2 \\
\underline{6xy-2y^2} \\
0
\end{array}
$$

65.
$$
\begin{array}{r}
x^2+x+1 \\
x-1\overline{\smash{\big)}\,x^3+0x^2+0x-1} \\
\underline{x^3-x^2} \\
0\ +x^2+0x \\
\underline{x^2-x} \\
0\ +\ x-1 \\
\underline{x-1} \\
0
\end{array}
$$

67. $x(2x-1)+8x^2$

$2x^2-x+8x^2$

$10x^2-x$

$10(-2)^2-(-2)$

$10\bullet4+2$

42

69. $3x^2-5y+2x^2+8y$

$5x^2+3y$

$5(4)^2+3(6)$

$5\bullet16+18$

98

71. $abc^3 + 2abc^3 + 5c^2$

$3abc^3 + 5c^2$

$3 \bullet 6 \bullet 7(12)^3 + 5(12)^3$

$218,448$

73. $1.8C + 32 - 0.09C + 1.6$

$1.71C + 33.6$

75. $2x - x^2 - 3x^2 + 6x$

$8x - 4x^2$

77. $36x - 4x^2$

79. $lh(1 + at)^2$

$lh(1 + at)(1 + at)$

$lh\left(1 + at + at + a^2t^2\right)$

$lh + 2atlh + a^2t^2lh$

81. $F = \dfrac{(6x + 2)(x + 5)^2}{3x + 1}$

$F = \dfrac{(6x + 2)(x + 5)(x + 5)}{3x + 1}$

$F = \dfrac{6x^3 + 62x^2 + 170x + 50}{3x + 1}$

$F = 2x^2 + 20x + 50$

$F = 2 \bullet 10^2 + 20 \bullet 10 + 50 = 450$

Chapter Test

1. $-10a$ and $9a$

2. $5ax^2$ and $-16ax^2$

3. $10a - 4b$

4. $-19a^2b$

5. $12a^2 - 18a^2b - 12ab$

6. $12a^3b^6$

7. $-8x^6y^3z^6$

8. $\dfrac{-4s^4}{ad}$

9. $6x^2 - 13xy - 28y^2$

10. $-xy - 3y^4 + 6xy^2$

11. $2x^2 + 4x + 2 + \dfrac{3}{2x - 1}$

12. 36

13. $2Vr - Va - Vb$

14. $4t - 4h - 2t^2 - 4th - 2h^2$

15. $\dfrac{r}{k} - \dfrac{2h^2}{kr} + \dfrac{h^2v^2}{k^2}$

Chapter 4
Simple Equations
and
Inequalities

4.1 Solving a Simple Equation

1. $x + 3 = 5$
   $x + 3 - 3 = 5 - 3$
   $x = 2$

3. $x + 5 - 5 = 8 - 5$
   $x = 3$

5. $\dfrac{2x}{2} = \dfrac{14}{2}$
   $x = 7$

7. $\dfrac{7x}{7} = 5(7)$
   $x = 35$

9. $\dfrac{6x}{6} = -3(6)$
   $x = -18$

11. $x + 6 - 6 = 2 - 6$
    $x = -4$

13. $2x - 5 + 5 = 13 + 5$
    $2x = 18$
    $\dfrac{2x}{2} = \dfrac{18}{2}$
    $x = 9$

15. $4x + 11 - 11 = 3 - 11$
    $4x = -8$
    $\dfrac{4x}{4} = \dfrac{-8}{4}$
    $x = -2$

17. $3 + 6x - 3 = 24 - x - 3$
    $6x = 21 - x$
    $6x + x = 21 - x + x$
    $7x = 21$
    $\dfrac{7x}{7} = \dfrac{21}{7}$
    $x = 3$

19. $3x - 6 + 6 = x - 18 + 6$
    $3x = x - 12$
    $3x - x = x - x - 12$
    $2x = -12$
    $\dfrac{2x}{2} = \dfrac{-12}{2}$
    $x = -6$

21. $2(x - 1) = x - 3$
    $2x - 2 = x - 3$
    $2x - 2 + 2 = x - 3 + 2$
    $2x = x - 1$
    $2x - x = -1$
    $x = -1$

23. $\left(\dfrac{x}{2}\right)2 = 2(x - 4)$
    $x = 2x - 8$
    $x - 2x = -8$
    $-x = -8$
    $x = 8$

25. $12\left(\dfrac{x}{3}+\dfrac{x}{4}\right)=12\left(\dfrac{7}{12}\right)$

$\quad 4x+3x=7$

$\quad\quad 7x=7$

$\quad\quad x=1$

27. $3(x-1)+x=2(x-1)$

$\quad 3x-3+x=2x-2$

$\quad\quad 4x-3=2x-2$

$\quad\quad 2x-3=-2$

$\quad\quad 2x=1$

$\quad\quad x=\dfrac{1}{2}$

29. $6x-1=3(x-2)+6$

$\quad 6x-1=3x-6+6$

$\quad 6x-1=3x$

$\quad 3x-1=0$

$\quad 3x=1$

$\quad x=\dfrac{1}{3}$

31. $0.2(3-x)-0.3(x-2)=0.1x$

$\quad 0.6-0.2x-0.3x+0.6=0.1x$

$\quad -0.5x+1.2=0.1x$

$\quad -0.6x=-1.2$

$\quad x=2$

33. $2x=18$

$\quad x=9$

35. $\dfrac{1}{2}x=16$

$\quad x=32$

37. $x+5=8$

$\quad x=3$

39. $\dfrac{x}{4}=17$

$\quad x=68$

41. $R=kl$

$\quad 75=k(2.6)$

$\quad 29=k$

$\quad R=29(75)$

$\quad R=2200\ \Omega$

43. $4x-5=\dfrac{1}{5}x$

$\quad 20x-25=x$

$\quad 19x=25$

$\quad x=\dfrac{25}{19}$

45. $x+5=x+7$; contradiction

$\quad x+5=x+5$; identity

47. a) $2x+2+1=3+2x$

$\quad 2x+3=3+2x$; identity

b) $3x+3=1+3x$; no

c) $5x=3x+4$; no

d) $2x-3=3x-3-x$

$\quad 2x-3=2x-3$; yes

e) $4x-5=4x-20$; no

49. $1.1=(T-76)\div40$

$\quad 44=T-76$

$\quad 120\ \ C=T$

51. $0.14n+0.06(2000-n)=0.09(2000)$

$\quad 0.14n+120-0.06n=180$

$\quad 0.08n=60$

$\quad n=750\ \text{gal}$

4.2 Simple Formulas and Literal Equations

1. $\dfrac{N}{A-s} = \dfrac{r(A-s)}{A-s}$

   $r = \dfrac{N}{A-s}$

3. $\dfrac{R_1 L_2}{L_1} = R_2$

5. $V_2 - at = V_1$

7. $\dfrac{PV}{R} = T$

9. $Id^2 = 5300CE$

   $\dfrac{Id^2}{5300E} = C$

11. $mRl = yd$

    $R = \dfrac{yd}{ml}$

13. $A = 180 - (B + C)$

    $A = 180 - B - C$

    $A - 180 + C = -B$

    $-A + 180 - C = B$

15. $Rm = CVL$

    $\dfrac{Rm}{CV} = L$

17. $PD_p = MD_m$

    $P = \dfrac{MD_m}{D_p}$

19. $L - 2d = 3.14(r_1 + r_2)$

    $\dfrac{L - 2d}{3.14} = r_1 + r_2$

    $\dfrac{L - 2d}{3.14} - r_1 = r_2$

21. $\dfrac{L}{L_o} = 1 + at$

    $\dfrac{L}{L_o} - 1 = at$

    $\dfrac{L}{L_o t} - \dfrac{1}{t} = a$

23. $\dfrac{p - p_a}{dg} = y_2 - y_1$

    $\dfrac{p - p_a}{dg} + y_1 = y_2$

25. $A(n_1 + n_2) = n_1 p_1 + n_2 p_2$

    $A(n_1 + n_2) - n_2 p_2 = n_1 p_1$

    $\dfrac{A(n_1 + n_2) - n_2 p_2}{n_1} = p_1$

27. $f(u + v_2) = f_s u$

    $u + v_2 = \dfrac{f_s u}{f}$

    $v_2 = \dfrac{f_s u}{f} - u$

29. $a - bc = d$

31. $ax = f - 3y$

    $x = \dfrac{f - 3y}{a}$

33. $x + y = \dfrac{3y}{2a}$

    $x = \dfrac{3y}{2a} - y$

35. $a = 2(b + 2)$

    $a = 2b + 4$

37. $x_1 - x_2 = a(3 + b)$

$$\frac{x_1 - x_2}{a} = 3 + b$$

$$\frac{x_1 - x_2}{a} - 3 = b$$

39. $2R_3 = R_1 + R_2$

$$2R_3 - R_1 = R_2$$

41. $7a(y + z) = 3(y + 2)$

$$y + z = \frac{3(y + 2)}{7a}$$

$$z = \frac{3(y + 2)}{7a} - y$$

43. $3(x + a) + a(x + y) = 4x$

$$a(x + y) = 4x - 3x - 3a$$

$$x + y = \frac{x - 3a}{a}$$

$$y = \frac{x - 3a}{a} - x$$

45. $A = \dfrac{a + b + c}{3}$

$$3A = a + b + c$$

$$3A - a - b = c$$

47. $I = \dfrac{E}{r + R}$

$$I(r + R) = E$$

$$r + R = \frac{E}{I}$$

$$R = \frac{E}{I} - r$$

49. $I = xr_1 + r_2(x + 1000)$

$$I - xr_1 = r_2(x + 1000)$$

$$\frac{I - xr_1}{x + 1000} = r_2$$

51. $x + 7y = C$

$$7y = C - x$$

$$y = \frac{C - x}{7}$$

53. $p = p_o + kh$

$$205 = 101 + 9.8h$$

$$104 = 9.8h$$

$$10.6 = h$$

$$10.6 \text{ m}$$

55. $V = \dfrac{1}{2}L(B + b)$

$$38.6 = \frac{1}{2}(16.1)(2.63 + b)$$

$$38.6 = 8.05(2.63 + b)$$

$$4.795031056 = 2.63 + b$$

$$2.17 = b$$

$$2.17 \text{ ft}^2$$

4.3 Simple Inequalities

1. $x - 7 + 7 > 5 + 7$

$x > 12$

3. $x + 2 < 7$

$x + 2 - 2 < 7 - 2$

$x < 5$

5. $\dfrac{x}{3} > 6$

$$\frac{x}{3}(3) > 6(3)$$

$x > 18$

7. $4x < -24$

$$\frac{4x}{4} < \frac{-24}{4}$$

$x < -6$

9. $-2x \geq -8$

$\dfrac{-2x}{-2} \leq \dfrac{-8}{-2}$

$x \leq 4$

11. $2x + 7 - 7 \leq 5 - 7$

$2x \leq -2$

$x \leq -1$

13. $2 - x - 2 > 6 - 2$

$-x > 4$

$x < -4$

15. $4x - 1 + 1 < x - 7 + 1$

$4x < x - 6$

$4x - x < -6$

$3x < -6$

$x < -2$

17. $3x - 6 + 6 > -9 + 6$

$3x > -3$

$x > -1$

19. $4x + 2 - 2 \leq 5x - 4 - 2$

$4x \leq 5x - 6$

$4x - 5x \leq -6$

$-x \leq -6$

$x \geq 6$

21. $2000 \leq M \leq 1,000,000$

23. $1550 + 60 = 1610$

$1550 - 60 = 1490$

$1490 \leq V \leq 1610$

25. $64.9 + 2.3 = 67.2$

$64.9 - 2.3 = 62.6$

$62.6 \leq p \leq 67.2$

27. $300 + 6x > 540$

$6x > 240$

$x > 40$

29. $1600(0.0025) = 4$

$3600(0.0025) = 9$

$4 \leq x \leq 9$

4.4 Problem Solving Strategies and Word Problems

1. Let x = the resistance in Ohms of one of the other resistors

$530 + x + x = 970$

$2x + 530 = 970$

$2x = 440$

$x = 220 \ \Omega$

3. Let x = the length of the original room. The new room would have

length $x - 2$

$p = 4s$

$56 = 4(x - 2)$

$56 = 4x - 8$

$64 = 4x$

$16 \ \text{ft} = x$

5. Let x = the value of the smaller piece. Then the larger
    piece would be 3x

    $x + 3x = 12$

    $4x = 12$

    $x = 3$ ft

    The larger piece is $3x = 3(3) = 9$ ft

7. Let x = the length of the longest side. Then the shorter sides
    length would be would be $x - 50$

    $x + x + x - 50 = 550$

    $3x - 50 = 550$

    $3x = 600$

    $x = 200$ m

    $200 - 50 = 150$ m

    200 m by 150 m

9. Let x = the storage capacity of the first computer. Then the second
    computer capacity would be would be 5.5x

    $5.5x + x = 65$

    $6.5x = 65$

    $x = 10$ Gbytes

    $5.5(10) = 55$ Gbytes

11. Let x = the cost of the competitors smoke detector. Then $4.29 + 2x$ is the
     cost of the more expensive alarm

     $4.29 + 2x + x = 87.36$

     $3x + 4.29 = 87.36$

     $3x = 83.07$

     $x = 27.69$

     $4.29 + 2(27.69) = \$59.67$

13. Let x = largest tank capacity, $\frac{x}{3}$ smallest tank capacity

     $\frac{x}{2}$ other tank capacity

     $x + \frac{x}{3} + \frac{x}{2} = 4400$

     $6x + 2x + 3x = 26400$

     $11x = 26400$

     $x = 2400$ gal largest capacity, 800 gal smallest capacity, and 1200 gal other capacity

15. Let x = how long they drive before they meet

     $50x + 55x = 420$

     $105x = 420$

     $x = 4$ h

17. Let x = speed of the faster missile, x − 600 speed of
    slower missle
    $5x + 5(x - 600) = 33000$
    $5x + 5x - 3000 = 33000$
    $10x = 36000$
    $x = 3600$ ft/s

19. Let x = pounds of 80% nickel alloy, 50 − x =
    pound of 40% nickel alloy
    $80x + 40(50 - x) = 50(50)$
    $80x + 2000 - 40x = 2500$
    $40x = 500$
    $x = 12.5$ lb of 80% nickel and 50 − x = 37.5 lb 40% nickel

21. Let x = how many tons of 80% crushed rock needed
    $80x + 30(150 - x) = 40(150)$
    $80x + 4500 - 30x = 6000$
    $50x = 1500$
    $x = 30$ tons

23. Let x = amount paid in state taxes
    $5x + x = 4800$
    $6x = 4800$
    $x = 800$ state; 4000 federal

25. Let x = how many quarts of 30% alcohol need to be drained off
    $30(8) - 30x + 100x = 50(8)$
    $240 + 70x = 400$
    $70x = 160$
    $x = 2.29$ qt

27. Let x = the side of the square
    The perimeter of the courtyard is 320 so
    $4x = 320$
    $x = 80$
    So the square walkway length is 80 + 6 = 86 so
    perimeter is 86(4) = 344 m

29. Let x = amount invested at 9%
    $8000(0.07) + 0.09x = 1550$
    $560 + 0.09x = 1550$
    $0.09x = 990$
    $x = \$11,000$

4.5 Ratio, Proportion and Variation

1. $\dfrac{12}{5}; \dfrac{3}{23}$

3. $\dfrac{21}{3} = \dfrac{7}{1}; \dfrac{2}{12} = \dfrac{1}{6}$

5. $\dfrac{6}{9} = \dfrac{2}{3}; \dfrac{12}{18} = \dfrac{2}{3}$

7. $\dfrac{6}{33} = \dfrac{2}{11}; \dfrac{8}{28} = \dfrac{2}{7}$

9. $\dfrac{30}{8} = \dfrac{15}{4}$

11. $\dfrac{9}{30} = \dfrac{3}{10}$

13. $\dfrac{8\text{ in}}{4\text{ ft}} = \dfrac{8\text{ in}}{48\text{ in}} = \dfrac{1}{6}$

15. $\dfrac{80\text{ s}}{3\text{ min}} = \dfrac{80\text{ s}}{180\text{ s}} = \dfrac{4}{9}$

17. $\dfrac{12\text{ m}}{6\text{ s}} = \dfrac{2\text{ m}}{1\text{ s}}$

19. $\dfrac{8}{36} = \dfrac{2\text{ lb}}{9\text{ ft}^3}$

21. $\dfrac{x}{2} = \dfrac{5}{8}$

    $8x = 10$

    $x = \dfrac{10}{8} = \dfrac{5}{4}$

23. $\dfrac{3}{14} = \dfrac{x}{4}$

    $14x = 12$

    $x = \dfrac{12}{14} = \dfrac{6}{7}$

25. $\dfrac{3}{x} = \dfrac{9}{15}$

    $9x = 45$

    $x = \dfrac{45}{9} = 5$

27. $\dfrac{4}{3} = \dfrac{12}{x}$

    $4x = 36$

    $x = 9$

29. $\dfrac{x}{6} = \dfrac{70}{30}$

    $30x = 420$

    $x = \dfrac{420}{30} = 14$

31. $\dfrac{x}{40} = \dfrac{7}{16}$

    $16x = 280$

    $x = \dfrac{280}{16} = \dfrac{35}{2}$

33. $\dfrac{6}{24} = \dfrac{1}{4}$

35. $\dfrac{908}{2} = \dfrac{x}{10}$

    $2x = 9080$

    $x = 4540\text{ g}$

37. $\dfrac{20}{15} = \dfrac{x}{25}$

    $15x = 500$

    $x = 33\dfrac{1}{3}\text{ in}$

39. $\dfrac{9}{5} = \dfrac{x}{280 - x}$

    $9(280 - x) = 5x$

    $2520 - 9x = 5x$

    $2520 = 14x$

    $x = 180\text{ mg and }100\text{ mg}$

41. $\dfrac{60}{8} = \dfrac{280}{x}$

    $60x = 2240$

    $x = 37\dfrac{1}{3}$ h

43. $\dfrac{3.5}{14} = \dfrac{x}{37}$

    $14x = 129.5$

    $x = 9.25$ mm

45. $\dfrac{27}{15} = \dfrac{9}{5}$

47. $y = kt$

49. $y = ks^2$

51. $t = \dfrac{k}{y}$

53. $y = kst$

55. $y = \dfrac{ks}{t}$

57. $x = \dfrac{kyz}{t^2}$

59. $y = ks$

    $25 = k(5)$

    $5 = k$

    $y = 5s$

61. $u = \dfrac{k}{d^2}$

    $17 = \dfrac{k}{4^2}$

    $272 = k$

    $u = \dfrac{272}{d^2}$

63. $y = \dfrac{kx}{t}$

    $6 = \dfrac{k(2)}{3}$

    $9 = k$

    $y = \dfrac{9x}{t}$

65. $s = kt$

    $20 = k(5)$

    $4 = k$

    $s = 4t$

    if t is 4 then s = 4(4) = 16

67. $q = \dfrac{k}{p}$

    $8 = \dfrac{k}{4}$

    $32 = k$

    $q = \dfrac{32}{p}$

    if p is 5 then $q = \dfrac{32}{5}$

69. $s = \dfrac{kp}{q^2}$

$100 = \dfrac{k(4)}{6^2}$

$900 = k$

$s = \dfrac{900p}{q^2}$

$s = \dfrac{900(75)}{5^2} = 2700$

71. $z = \dfrac{k}{xy}$

$4 = \dfrac{k}{2(3)}$

$24 = k$

$z = \dfrac{24}{xy}$

$z = \dfrac{24}{12(4)} = \dfrac{1}{2}$

73. $E = kI$

$115 = k(5)$

$23 = k$

$E = 23I$

75. $F = \dfrac{kQ_1Q_2}{s^2}$

77. $H = ks^3$

$10.5 = k(10)^3$

$0.0105 = k$

$H = 0.0105(15)^3 = 35.4 \text{ hp}$

79. $R = \dfrac{kl}{d^2}$

$6.5 = \dfrac{k(100)}{(0.00200)^2}$

$0.00000026 = k$

$R = \dfrac{0.00000026(25)}{(0.00750)^2} = 0.116 \ \Omega$

81. $\dfrac{150}{200} = \dfrac{180}{x}$

$150x = 36000$

$x = 240 \text{ r/min}$

Review Exercises

1. $x - 3 + 3 = 5 + 3$

$x = 8$

3. $\dfrac{3y}{3} = \dfrac{27}{3}$

$y = 9$

5. $4x + 21 - 21 = 5 - 21$

$4x = -16$

$x = -4$

7. $3x + 3 = x + 11$

$2x + 3 = 11$

$2x = 8$

$x = 4$

9. $2 - 4t = 11 - t$

$2 = 11 + 3t$

$-9 = 3t$

$t = -3$

11. $7s - 7 + 2s + 4 = 3s$

$9s - 3 = 3s$

$6s - 3 = 0$

$6s = 3$

$s = \dfrac{1}{2}$

13. $R = R_1 + R_2 + R_3$

$R - R_1 - R_2 = R_3$

15. $rs_2 = \dfrac{ms_1}{s_2} s_2$

$rs_2 = ms_1$

$\dfrac{rs_2}{r} = \dfrac{ms_1}{r}$

$s_2 = \dfrac{ms_1}{r}$

17. $\dfrac{d_m}{A} = \dfrac{(n-1)}{A} A$

$\dfrac{d_m}{A} = n - 1$

$\dfrac{d_m}{A} + 1 = n$

19. $M_1V_1 + M_2V_2 = PT$

$M_1V_1 = PT - M_2V_2$

$M_1 = \dfrac{PT - M_2V_2}{V_1}$

21. $RH = \dfrac{wL}{w + L}$

$H = \dfrac{wL}{R(w + L)}$

23. $W + H_1 - H_2 = T(S_1 - S_2)$

$\dfrac{W + H_1 - H_2}{S_1 - S_2} = T$

25. $a(2 + 3x) - 2ax = 3y$

$2a + 3ax - 2ax = 3y$

$2a + ax = 3y$

$\dfrac{2a + ax}{3} = y$

27. $a(x + b) = b(x + c)$

$\dfrac{a(x + b)}{b} = x + c$

$\dfrac{a(x + b)}{b} - x = c$

29. $6 - 3x + 2x = a(a + b)$

$6 - x = a(a + b)$

$\dfrac{6 - x}{a} = a + b$

$\dfrac{6 - x}{a} - a = b$

31. $3a(a + 2x) + a^2 = a(2 + a)$

$3a^2 + 6ax + a^2 = 2a + a^2$

$4a^2 + 6ax = 2a + a^2$

$6ax = 2a - 3a^2$

$x = \dfrac{2a - 3a^2}{6a} = \dfrac{1}{3} - \dfrac{a}{2}$

33. $x - 2 < 7$

$x < 9$

35. $2x \geq 10$

$x \geq 5$

37. $3x + 10 < 1$

$3x < -9$

$x < -3$

39. $6 - x > 10$

$-x > 4$

$x < -4$

41. $5x + 5 < x - 3$

$4x + 5 < -3$

$4x < -8$

$x < -2$

43. $12 - 2x \geq 3x - 3$

$15 - 2x \geq 3x$

$15 \geq 5x$

$3 \geq x$

45. $\dfrac{2 \text{ ft}}{36 \text{ in}} = \dfrac{24 \text{ in}}{36 \text{ in}} = \dfrac{2}{3}$

47. $\dfrac{4 \text{ min}}{40 \text{ s}} = \dfrac{240 \text{ s}}{40 \text{ s}} = \dfrac{6}{1}$

49. $\dfrac{x}{3} = \dfrac{8}{9}$

$9x = 24$

$x = \dfrac{24}{9} = \dfrac{8}{3}$

51. $\dfrac{3}{10} = \dfrac{x}{15}$

$10x = 45$

$x = \dfrac{45}{10} = \dfrac{9}{2}$

53. $2x + 1 = 9$

$2x = 8$

$x = 4$

55. $3x = 8 + x$

$2x = 8$

$x = 4$

57. $\dfrac{x}{5} = \dfrac{7}{15}$

$15x = 35$

$x = \dfrac{35}{15} = \dfrac{7}{3}$

59. $\dfrac{x}{12} = \dfrac{7}{8}$

$8x = 84$

$x = \dfrac{84}{8} = \dfrac{21}{2}$

61. $y = kx$

$24 = k(4)$

$6 = k$

$y = 6x$

63. $m = \dfrac{k}{\sqrt{r}}$

$5 = \dfrac{k}{\sqrt{9}}$

$5 = \dfrac{k}{3}$

$15 = k$

$m = \dfrac{15}{\sqrt{r}}$

65. $f = \dfrac{km}{p}$

$8 = \dfrac{k(4)}{5}$

$40 = k(4)$

$10 = k$

$f = \dfrac{10(3)}{6} = 5$

67. $s = \dfrac{ktu}{v}$

$27 = \dfrac{k(3)(5)}{6}$

$162 = k(15)$

$10.8 = k$

$s = \dfrac{10.8(2)(3)}{4} = 16.2$

69. Let $x =$ the height then width is $x - 18$

$2(x - 18) + 2x = 180$

$2x - 36 + 2x = 180$

$4x - 36 = 180$

$4x = 216$

$x = 54$ in; width is 36 in

71. Let x = the cost of manufacturing one of the second storage devices,
    then 3x is the cost of one of the first storage devices

    $3x + 2(3x) = 450$

    $9x = 450$

    $x = \$50$ and $\$150$

73. Let x = score on 4th test

    $$\frac{68 + 73 + 84 + x}{4} = 80$$

    $$\frac{225 + x}{4} = 80$$

    $225 + x = 320$

    $x = 95$

75. Let x = amount in second account.

    $x + 3000 + x + x - 4500 = 21600$

    $3x - 1500 = 21600$

    $3x = 23100$

    $x = \$7700; \ \$10{,}700; \ \$3200$

77. Let x = the speed of one jet.

    $3x + 3(x + 300) = 12000$

    $3x + 3x + 900 = 12000$

    $6x + 900 = 12000$

    $6x = 11100$

    $x = 1850$ km/h; 2150 km/h

79. Let x = how much water must be added

    $0x + 15(100) = 60(x + 15)$

    $1500 = 60x + 900$

    $600 = 60x$

    $x = 10$ L

81. Let x = amount added

    $0.08x + 0.08(12500) = 1720$

    $0.08x + 1000 = 1720$

    $0.08x = 720$

    $x = \$9000$

83. $w = kx$

    $8.0 = k(2.3)$

    $3.5 = k$

    $w = 3.5(5.7)$

    $w = 20.0$ lb

Chapter Test

1. $2x = -9$

    $$x = \frac{-9}{2}$$

2. $8(n - 4) = 15$

    $8n - 32 = 15$

    $8n = 47$

    $$n = \frac{47}{8}$$

3. $3t - 14 + 2t = 10t + 5$

    $5t - 14 = 10t + 5$

    $-19 = 5t$

    $$t = \frac{-19}{5}$$

4. $$\frac{R}{n^2} = Z$$

5. $\dfrac{V - I_r}{I} = R$

6. $J + 1 = \dfrac{f}{2B}$

$J = \dfrac{f}{2B} - 1$

7. $2x > 12$

$x > 6$

8. $-7x - 7 < 2x + 8$

$-15 < 9x$

$\dfrac{-15}{9} < x$

$\dfrac{-5}{3} < x$

9. $\dfrac{19}{3}$

10. $\dfrac{6 \text{ min}}{45 \text{ s}} = \dfrac{360}{45} = \dfrac{8}{1}$

11. $600 = 8x$

$75 = x$

12. $x + 129 + x = 390$

$2x = 261$

$x = \$130.50 \text{ and } \$259.50$

13. $\dfrac{5}{1.7} = \dfrac{27}{x}$

$5x = 45.9$

$x = 9.18 \text{ lb}$

14. $p = kt$

$60000 = k(2)$

$30000 = k$

$p = 30000t$

15. $R = kl$

$75 = k(2.6)$

$29 = k$

$R = 29(75)$

$R = 2200 \ \Omega$

Chapter 5
Graphs

5.1 Functions and Function Notation

1. Dependent y, Independent x

3. Dependent p, Independent v

5. Multiplication

7. Square the value of the independent variable

9. $f(x) = 5 - x$

11. $f(t) = t^2 - 3t$

13. $f(0) = 0; f(3) = 3$

15. $f(4) = 2(4) - 1 = 7$
$f(-2) = 2(-2) - 1 = -5$

17. $f(0) = 3(0) - 2 = -2$
$f(\frac{1}{2}) = 3\left(\frac{1}{2}\right)^2 - 2 = \frac{3}{4} - 2 = \frac{3}{4} - \frac{8}{4} = \frac{-5}{4}$

19. $f(2) = 3 - 2^2 = 3 - 4 = -1$
$f(-0.3) = 3 - (-0.3)^2 = 3 - 0.09 = 2.91$

21. $f(-1) = (-1)^3 = -1$
$f(2) = 2^3 = 8$

23. $f(1) = 3(1) - 1^3 = 3 - 1 = 2$
$f(-2) = 3(-2) - (-2)^3 = -6 + 8 = 2$

25. $f(-3) = \frac{-3}{-3-3} = \frac{-3}{-6} = \frac{1}{2}$
$f(3) = \frac{3}{3-3} = \frac{3}{0} = $ undefined (cannot divide by 0)

27. $f(a^2) = a^2 - 2\left(a^2\right)^2 = a^2 - 2a^4$
$f(\frac{1}{a}) = \frac{1}{a} - 2\left(\frac{1}{a}\right)^2 = \frac{1}{a} - \frac{2}{a^2} = \frac{a-2}{a^2}$

29. Not a function

31. Not a function

33. Answers will vary

35. $x \geq -7$

37. x can be any real number except 0

39. $v(e) = e^3$

41. $C(P) = P + 0.05P$

43. $V(i) = 5i$

45. a) $C(h) = \{12 \text{ for } 0 < h \le 6; 12 + 6(h - 6) \text{ for } h > 6$

  b) Independent h; Dependent C

  c) h > 0

5.2 Rectangular Coordinate System

1. $A(2,1); B(-2,3)$

3. $E(4,0); F(-2,1)$

5. $I(1,5.5); J(3,5.5)$

7. $M(-9.5,0); N(-9.5,2)$

9.

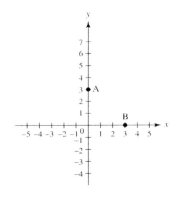

11.

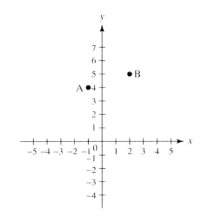

13.

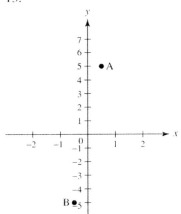

15.

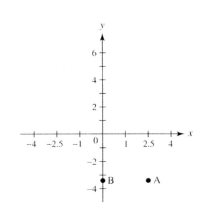

17. I, II

19. III, II

21. 0

23. I and III

25. I

5.3 The Graph of a Function

1.

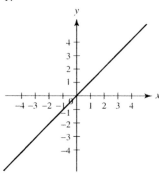

3.

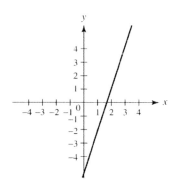

5.

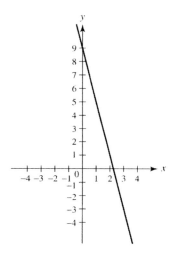

7.

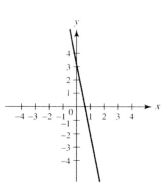

9.

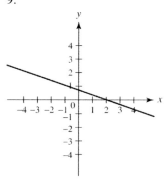

11.

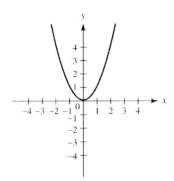

13.

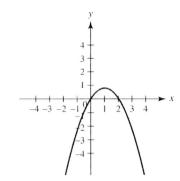

15.

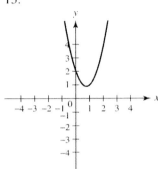

17.

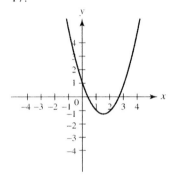

19.

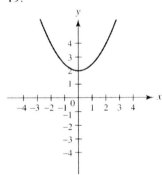

21.

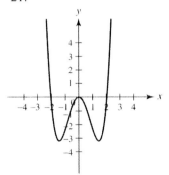

23.

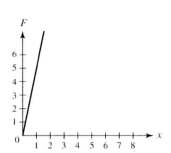

25.

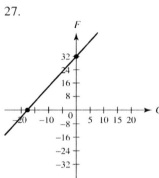

27.

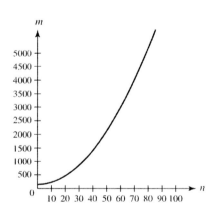

29.

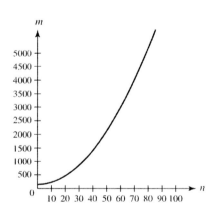

31.

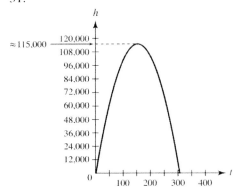

33.

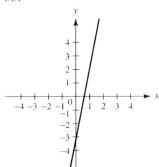

35.

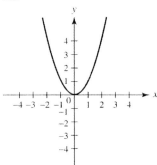

5.4 Graph of a Linear Function

1. $\dfrac{6}{3} = 2$

3. $\dfrac{-5}{5} = -1$

5. $m = \dfrac{7-4}{5-2} = \dfrac{3}{3} = 1$

7. $m = \dfrac{-6-5}{5-(-2)} = \dfrac{-11}{7}$

9. $m = \dfrac{-3.4-(-3)}{0-2.5} = \dfrac{-0.4}{-2.5} = \dfrac{4}{25}$

11. $m = \dfrac{7-3}{1-1} = \dfrac{4}{0} = \text{undefined}$

13. $m = \dfrac{4-4}{-2-5} = \dfrac{0}{-7} = 0$

15. $m = \dfrac{0.2-0.5}{-0.2-0.4} = \dfrac{-0.3}{-0.6} = \dfrac{3}{6} = \dfrac{1}{2}$

17. $0 + 2y = 4$

   $2y = 4$

   $y = 2$

   $x + 2(0) = 4$

   $x = 4$

   $(0, 2); (4, 0)$

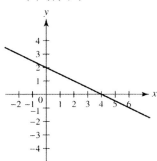

19. $0 - 2y = 20$

   $y = -10$

   $5x - 0 = 20$

   $x = 4$

   $(0, -10); (4, 0)$

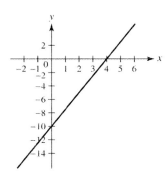

21. $y = 4.5$

   $0 = 0.25x + 4.5$

   $-4.5 = 0.25x$

   $-18 = x$

   $(0, 4.5); (-18, 0)$

23. $y = 2$

   $0 = \frac{5}{4}x + 2$

   $-2 = \frac{5}{4}x$

   $\frac{-8}{5} = x$

   $(0, 2); (\frac{-8}{5}, 0)$

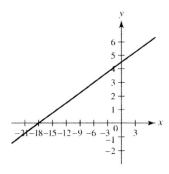

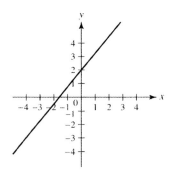

25. $m = -2$

y-intercept $= (0,1)$

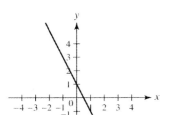

27. $m = 1$

y-intercept $= (0,4)$

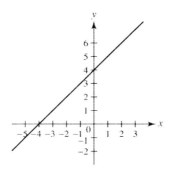

29. $y = \dfrac{4}{-2}x - \dfrac{7}{2}$

$m = -2$

y-intercept $= \left(0, \dfrac{-7}{2}\right)$

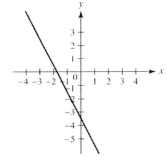

31. $y = -\dfrac{5}{8}x + \dfrac{5}{4}$

$m = -\dfrac{5}{8}$

y-intercept $= \left(0, \dfrac{5}{4}\right)$

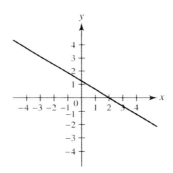

5.5 Graphing Non-Linear Functions

1.

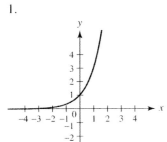

3.

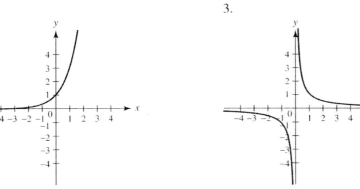

5.

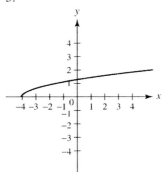

7.

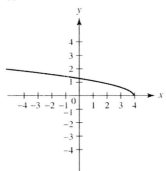

9.

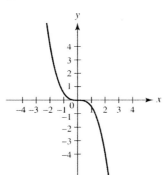

11.

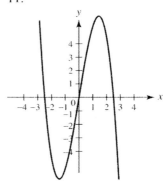

13.

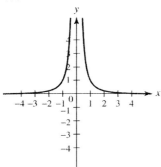

15.

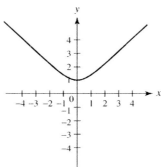

17.

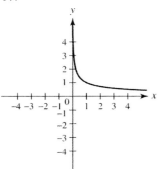

19.

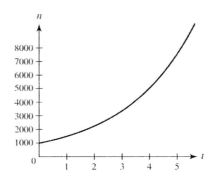

21.

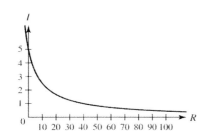

23.

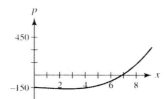

25.

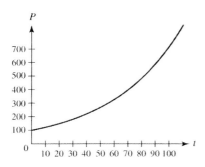

5.6 Graphical Solutions

1. $1.4, -1.2$

3. $9, -8$

5. $6.5, 11.6$

7. $9.2, 1.7$

9. $8.9, 2.1$

11. $-2.1, 5.0$

13. 0.7

15. $-1.2, 1.7$

17. 1.7

19. 4.2

21. 3.5

23. 1.5

25. 0.0055 C, 0.0001 C

27. 35 g, 70 g

29. $0.8F = 22$

$$F = \frac{22}{0.8} = \frac{220}{8} = \frac{110}{4} = \frac{55}{2} = 28 \text{ lb}$$

31. $2\pi r^3 - 57.8 = 0$

$2\pi r^3 = 57.8$

$r^3 = 9.1991557$

$\sqrt[3]{r^3} = \sqrt[3]{9.1991557}$

$r = 2.10$ in

5.7 Graphing Inequalities

1.

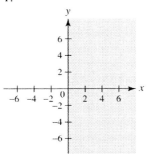

3.

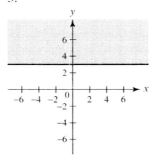

5.

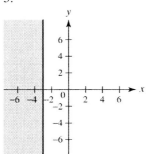

7.

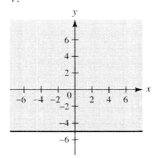

9.

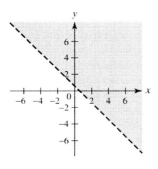

11.

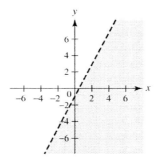

13.

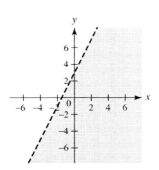

15.

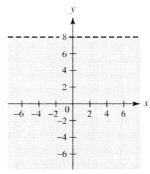

17.

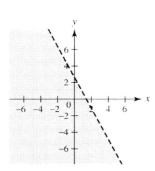

19.

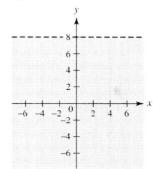

21.

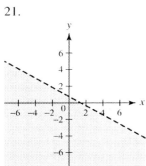

23.

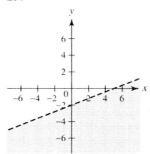

25.

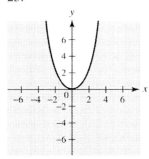

27.

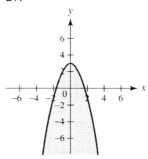

29.

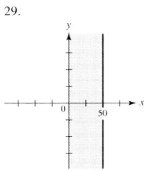

31.

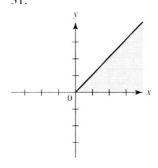

Review Exercises

1. $f(0) = 0 + 3 = 3$
   $f(-1) = -1 + 3 = 2$

3. $f(-2) = 2 - (-2) = 4$
   $f\left(\dfrac{1}{3}\right) = 2 - \dfrac{1}{3} = \dfrac{5}{3}$

5. $f\left(\sqrt{2}\right) = 2\left(\sqrt{2}\right)^2 - 1 = 4 - 1 = 3$
   $f\left(\dfrac{1}{2}\right) = 2\left(\dfrac{1}{2}\right)^2 - 1 = \dfrac{2}{4} - 1 = \dfrac{-1}{2}$

7. $f(-3) = (-3^2) + 2(-3) - 3 = 9 - 6 - 3 = 12$
   $f(0.2) = (0.2)^2 - 2(0.2) - 3 = 0.04 - 0.4 - 3 = -3.36$

9. $f(0) = 0$
   $f(-2) = -(-2)^3 = 8$

11. $f(0) = \sqrt{4(0)+1} = \sqrt{1} = 1$
    $f(6) = \sqrt{4(6)+1} = \sqrt{25} = 5$

13.-20.

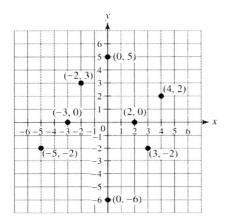

21.

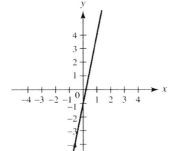

23.

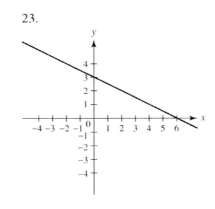

25.

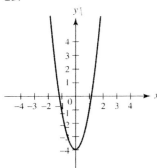

27.

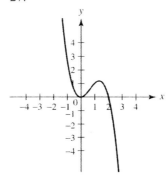

29.

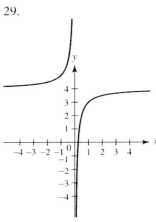

31.

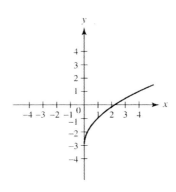

33. $m = \dfrac{7-6}{-5-2} = \dfrac{1}{-7}$

35. $m = \dfrac{-6-5}{5-(-2)} = \dfrac{-11}{7}$

37. $m = \dfrac{12-(-13)}{0-5} = \dfrac{25}{-5} = -5$

39. $(0,8),(4,0)$

41. $(0,-6),\left(\dfrac{9}{2},0\right)$

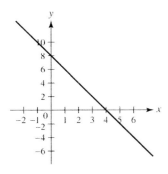

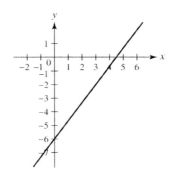

43. $(0, 6.5), (-13, 0)$

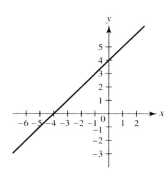

45. $m = -2$

y-intercept = $(0, 0)$

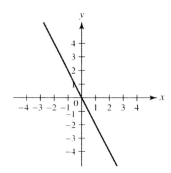

47. $y = x + 4$

$m = 1$

y-intercept = $(0, 4)$

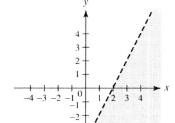

49.

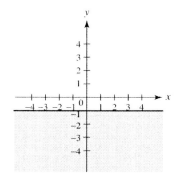

51.

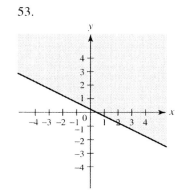

53.

55.

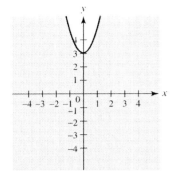

57.

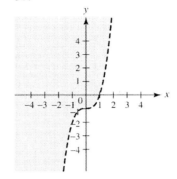

59.

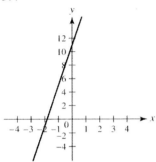

61.

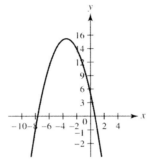

63. $T(C) = 0.06C$

65. $H(I) = 240I^2$

67. $700, $1700, $2400

69. $L = 100\left(1 + 0.0003(5)^2\right) = 100 \text{ ft}^2$

$L = 100\left(1 + 0.0003(10)^2\right) = 103 \text{ ft}^2$

$L = 100\left(1 + 0.0003(20)^2\right) = 112 \text{ ft}^2$

71. $F(C) = \dfrac{9}{5}C + 32$

$F = C$ at -40

Chapter Test

1. a) $-7, 2, -10$
   b) $33, 26$
   c) 18

2. (a)

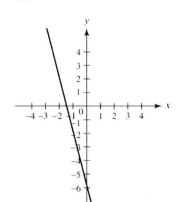

2. (b)

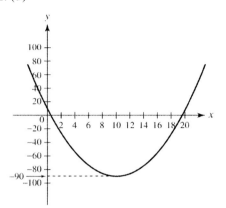

3. $m = \dfrac{1 - (-4)}{-7 - 3} = \dfrac{5}{-10} = \dfrac{1}{-2}$

4. $(2, 0), (0, -4)$

5. $y = \dfrac{4}{3}x + 2$

   $m = \dfrac{4}{3}$

   y-intercept $= (0, 2)$

6. (a)

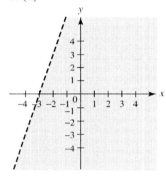

6. (b)

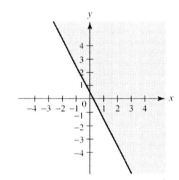

7. (a) -1.5 (b) $-5.16, 1.16$

8. (a)

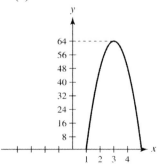

8. (b) 1 s, 5 s   8. (c) 64 ft

9. a) $900, $4400, $3000
   b) 53 items

Chapter 6
Introduction to Geometry

6.1 Basic Geometric Figures

1. estimates will vary; 24°

3. estimates will vary; 102°

5. 56° 24'; multiply 0.4 by 60'

7. 136° 27'; multiply 0.45 by 60'

9. 156.25°; divide 15' by 60

11. 67.1°; divide 6' by 60

13. ∠EBD, ∠DBC; acute angles are less than 90°

15. ∠ABD; obtuse angles are greater than 90°

17. ABE, EBC; right angles are formed by perpendicular lines

19. 115°; 180 − 65 = 115

21. ΔBDA; scalene triangles have three unequal sides

23. 54°; since ΔCDB is isosceles, the base angles are equal

25. 5 in; the hypotenuse is the longest side

27. A and C, B and D; opposite angles do not share a side

29. 8 cm; opposite sides in a parallelogram are equal

31. 3 in; all sides of a rhombus are equal

33. 4 cm; the radius is half the diameter

35.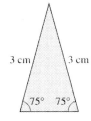

37.

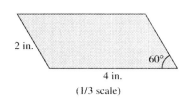

4 in.
(1/3 scale)

39.

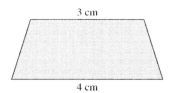

41.

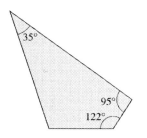

43a. Yes, if a triangle has three equal sides, then it has two

43b. No, a triangle may have two equal sides without having three

45. A rhombus with interior angles 60° and 120°; equilateral triangles have angles of 60°

47. 54°; the angles are complementary

49. 94 km; the diameter is twice the radius

51. west; west is 90° to the left of north

6.2 Perimeter

1. 12 ft; $3 + 4 + 5 = 12$

3. 123 mm; $18 + 34 + 25 + 46 = 123$

5. 896 m; $320 + 278 + 298 = 896$

7. 84.8 in; $a + b + c + d = 84.8$

9. 193.8 mm; $s + s + s = 193.8$

11. 57.1 in; $2(15.3) + 26.5 = 57.1$

13. 2.6 m; $4(0.65) = 2.6$

15. 0.6 mi; $4(0.15) = 0.6$

17. 168 cm; $2a + 2b = 168$

19. 230.6 ft; $2l + 2w = 230.6$

21. 30.2π cm ≈ 94.88 cm
    $2\pi r \approx 94.88$

23. 6.74π in ≈ 21.17 in
    $\pi d \approx 21.17$

25. 36 ft; $4 + 5 + 8 + 12 + 7 = 36$

27. 23 cm;
    $6.60 + 2.00 + 6.60 + 2(2.90) + 2.00$

29. 45.36 in;
    $3(9.75) + 3.18 + 9.75 + 3.18$

31. $15.9 + 2.65\pi$ m ≈ 24.23 m;
    $3(5.30) + 0.5(5.30\pi)$

33. $P = s + s + a = 2s + a$

35. $P = 2r + \pi r$;
    $P = 2r + 0.5(\pi \times 2r)$

37. $P = 5s$

39. $P = 2a + b_1 + b_2$

41. \$69.30; $1.65[2(9) + 2(12)]$

43. 7920π mi $\approx 24{,}881.41$ mi;
    $C = 2\pi \times 3960$

45. 891.27 ft; $d = 2800/\pi$

47. 12.57 ft; $4[0.25(\pi \times 4)]$

49. 55.13 in; $2(15) + 2[0.5(2\pi \times 4)]$

51. 1428.50 ft;
    $205 + 110 + 85 + 170 + 85 + 110 + 205 + 110 + 50 + 0.5(\pi \times 120) + 110$

53. 26.00 in; $81.68/\pi$

55. 8.89 cm;
    $20{,}109 \div 360 = 55.86 = C$,
    $C / 2\pi = r = 8.89$

57. 1.46π cm ≈ 4.59 cm;
    $0.558\pi + 0.902\pi$

6.3 Area

1. 2700 cm^2; 60×45

3. 168 in^2; 24×7.0

5. 57.76 in^2; 7.60^2

7. 6.4516 in^2; 2.54^2

9. 2448 mm^2; 72.0 × 34.0

11. 24.38 cm^2; 5.30 × 4.60

13. 105.6 in^2; 16.5 × 6.40

15. 40 ft^2; 8 × 5

17. 0.48 m^2; 0.750 × 0.640

19. 672 cm^2; 42 × 16

21. 475.2 ft^2; 16.5 × 28.8

23. 31,365 cm^2; 153 × 205

25. 129.91 cm^2; $s = 29.45$, $A = \sqrt{29.45(29.45 - 21.2)(29.45 - 12.3)(29.45 - 25.4)}$

27. 12.05 ft^2; $s = 8.15$, $A = \sqrt{8.15(8.15 - 5.2)(8.15 - 6.4)(8.15 - 4.7)}$

29. 0.000258 km^2; $\frac{1}{2}(0.0120)(0.0250 + 0.0180)$

31. 31.02 m^2; $\frac{1}{2}(6.33)(4.55 + 5.25)$

33. 0.72 ft^2; $\pi(0.478)^2$

35. 1692.39 ft^2; $\pi(23.21)^2$

37. 21.65 in^2; $\pi(2.625)^2$

39. 1.27 ft^2; $r = 4/2\pi$, $A = \pi r^2$

41a. 36 in; 2(12) + 2(6.0)

41b. 60 in^2; (5.0)(12)

43a. 60 cm; 4(15)

43b. 225 cm^2; (15)^2

45a. 30 in; 5.0 + 12 + 13

45b. 30 in^2; $\frac{1}{2}(5.0)(12)$

47a. 26.6 in; 3(5.20) + 11.0

47b. 34.99 in^2; $\frac{1}{2}(4.32)(5.20 + 11.0)$

49. $179.20; (16.0)(32.0)($0.35)

51. 1.94 gallons; [2(12)(8) + 2(16)(8) + 12(16) − 3(2)(3) − 2(3)(6.5)]/300

53. 122,040 m^2; $\frac{1}{2}(568 + 110)(360)$

55. 665.08 cm^2; $\pi[(38.0 - 8.90)/2]^2$

57. $12,901.58; $\pi(2.34/2)^2($3000)

59. The area is multiplied by 9;
$A = (3l)(3w) = 9lw$

61a. 203 ft^2; $(14)(12) + \frac{1}{2}(5)(14)$

61b. $473.67; $203 \div 9 \times 21$

6.4 Volume

1. 20,089.6 mm^3; $(73.0)(17.2)(16.0)$

3. 0.08 cm^3; $(0.87)(0.61)(0.15)$

5. 4608 m^3; $(16)(16)(18)$

7. 1,530,000 cm^3; $(120)(85)(150)$

9. 3375 cm^3; 15^3

11. 0.512 in^3; $(0.800)^3$

13. 0.011 m^3; $(0.22)^3$

15. 7155.58 cm^3; $(19.27)^3$

17. 18,849.56 cm^3; $\pi(20.0)^2(15.0)$

19. 2,544.69 ft^3; $\pi(15.0)^2(3.60)$

21. 58,916,925.97 mm^3; $\pi(366)^2(140)$

23. 54,000 ft^3; $\frac{1}{3}(3600)(45.0)$

25. 958.33 mm^3; $\frac{1}{3}(25.0)^2(4.60)$

27. 1466.08 ft^3; $\frac{1}{3}\pi(10.0)^2(14.0)$

29. 27,154.62 cm^3; $\frac{1}{3}\pi\left(\frac{62.8}{2}\right)^2(26.3)$

31. 85,226.87 cm^3; $\frac{4}{3}\pi(27.3)^3$

33. 14,710.23 m^3; $\frac{4}{3}\pi(15.2)^3$

35. 20,579.53 in^3; $\frac{4}{3}\pi\left(\frac{34}{2}\right)^3$

37. 0.736 mm^3; $\frac{4}{3}\pi\left(\frac{1.12}{2}\right)^3$

39. 1728 in^3; 12^3

41. 1,000,000 cm^3; 100^3

43. 1344 ft^3; $(12)(14)(8.0)$

45. 15.86 yd^3; $(2.32)(1.28)(5.34)$

47. No; $(9.0)(12)(8.0) = 864 \text{ ft}^3, (4)(7)(1.8)(30,000) \div 12^3 = 875 \text{ ft}^3$

49. 122,674.58 bu; $\pi(24.3)^2(82.0) \div 1.24$

51. 3,333,333,33 yd^3; $\dfrac{1}{3}(250)^2(160)$

53. 23,524.25 lb; $\dfrac{1}{3}\pi(6.0)^2(10.0)\times 62.4$

55. 4.71 m^3; $\dfrac{4}{3}\pi\left(\dfrac{2.08}{2}\right)^3$

57. 99.54 in^3; $\dfrac{4}{3}\pi\left(\dfrac{8.50-2.75}{2}\right)^3$

Review Exercises for Chapter 6

1. 37° 30'; 0.5×60

3. 12° 33'; 0.55×60

5. 63.5°; $30\div 60$

7. 105.9°; $54\div 60$

9. 40.2 in; $17.5+13.8+8.9$

11. 25.5 mm; $3(25.5)$

13. 27.2 m; $4(6.8)$

15. 1798 ft; $2(692)+2(207)$

17. 278 cm; $2(96)+2(43)$

19. 26.70 ft; $2\pi(4.25)$

21. 153 mm; $2(42) + 23 + 2(23)$

23. 15.80 mm;
    $2.38+2(4.84)+0.5(\pi)(2.38)$

25. 224 cm^2; $(12.8)(17.5)$

27. 3.08 yd^2; $(2.68)(1.15)$

29. 4.80 cm^2; $0.5(4.68)(2.05)$

31. 3.06 in^2; $0.5(1.88)(3.25)$

33. 0.00148 km^2;
    $0.5(0.016)(0.067+0.118)$

35. 35.47 in^2; $\pi(3.36)^2$

37. 138,544.24 mm^2; $\pi\left(\dfrac{420}{2}\right)^2$

39. 3.30 m^3; $s = 4.14$, $A = \sqrt{4.14(4.14-2.76)^3}$

41. 3.6 m^3; $(2.00)(1.50)(1.20)$

43. 42.875 yd^3; $(3.50)^3$

45. 678.08 cm^3; $\frac{4}{3}\pi(5.45)^3$

47. 27.39 yd^3; $\frac{4}{3}\pi\left(\frac{3.74}{2}\right)^3$

49a. 140 ft; $2(22)+2(48)$

49b. 204 ft; $2(22+16)+2(48+16)$

51. $30,729.08 \text{ lb}$; $\frac{4}{3}\pi(2.432)^3 \times 510$

53.

$$\text{Let } x = \text{diameter}$$
$$\text{Then } x + 1088 = \text{circumference}$$
$$x + 1088 = \pi x$$
$$1088 = \pi x - x = x(\pi - 1)$$
$$x = \frac{1088}{\pi - 1} = 508.03 \text{ mm}$$

55. $128,000 \text{ lb}$; $(32.0)(16.0)(4.00)(62.5)$

57. 5.17 ft^2; $(3.60)(1.00)+0.5(\pi)(1.00)^2$

59. 146.14 gal; $(2.25)(3.75)(8)(500)(0.00433)$

Chapter Six Test

1. $45° \, 42'$; 0.7×60

2. $62.58°$; $35 \div 60$

3a. 80.4 in; $35 + 27.6 + 17.8$

3b. 79.65 cm; $11.2 + 34.25 + 18.7 + 15.5$

3c. 48.69 in; $2\pi(7.75)$

4a. 3.06 ft^2; $0.5(3.25)(1.88)$

4b. 7604.66 mm^2; $\pi\left(\frac{98.4}{2}\right)^2$

4c. 13.75 yd^2; $(2.5)(5.5)$

5a. 9771.61 in^3; $\pi(36.0)^2(2.40)$

5b. $160,416.67 \text{ ft}^3$; $\frac{1}{3}(3850)(125)$

5c. 5651.65 mm^3; $\frac{4}{3}\pi\left(\frac{22.1}{2}\right)^3$

6. 12 cm; $6(2)$

7. $26,201 \text{ mi}$; $2\pi\left(\frac{7920}{2}+210\right)$

8. 30.43 ft^2; $(4.0)(8.0)-2\left[\pi\left(\frac{1}{2}\right)^2\right]$

9. $49,762,828 \text{ ft}^3$; $\pi\left(\frac{4}{2}\right)^2(750)(5280)$

10. $2,352,071 \text{ ft}^3$; $\frac{4}{3}\pi\left(\frac{165}{2}\right)^3$

Chapter 7
Simultaneous Linear Equations

7.1 Graphical Solution of Two Simultaneous Equations

1. $x = 3, y = 0$

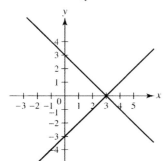

3. $x = 2, y = 1$

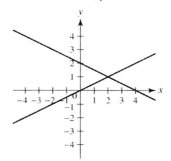

5. $r = 4, x = -3$

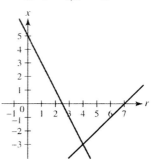

7. $x = 2, y = 2$

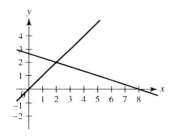

9. $x = -1, y = 2$

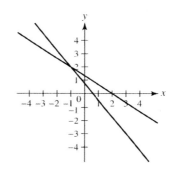

11. $x = -2, y = 1.5$

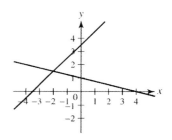

13. $a = -1.9, b = -0.5$

15. inconsistent system

17. $p = 2.4, q = 1.1$

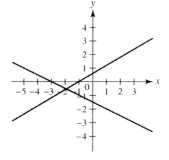

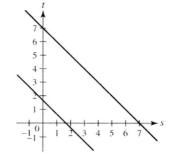

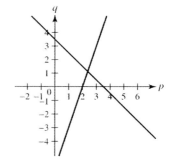

19. $x = 1.9, y = -2.2$

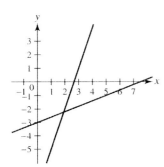

21. $x = 5.9, y = -0.2$

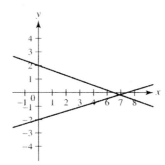

23. $x = 0.8, y = 3.6$

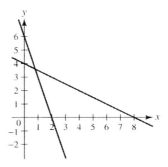

25. inconsistent

27. dependent

29. inconsistent

31. inconsistent

33. $x = 95, y = 25$

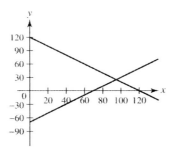

35. $d = 4, g = 6$

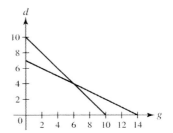

37. $T_1 = 12.3, T_2 = 14.3$

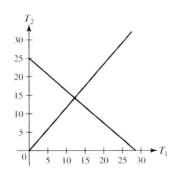

39. $x = 8.5, y = 11$

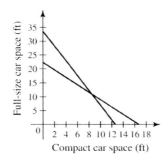

7.2 Substitution Method

1. $x = 5, y = 8;$
$$(y - 3) + y = 13$$
$$2y = 16$$
$$y = 8$$

3. $x = 6, y = 4;$
$$2(y + 2) - y = 8$$
$$2y + 4 - y = 8$$
$$y = 4$$

5. $x = 0.4, y = 0.6;$
$$5(1 - y) + 10y = 8$$
$$5 - 5y + 10y = 8$$
$$5y = 3$$
$$y = 0.6$$

7. $x = -1, y = 2;$
$$3x + 2(-2x) = 1$$
$$3x - 4x = 1$$
$$-x = 1$$
$$x = -1$$

9. $x = 2, y = 1;$
$$2(2y) - 3y = 1$$
$$4y - 3y = 1$$
$$y = 1$$

11. $x = 0, y = 0;$
$$3(4y) = 2y$$
$$12y = 2y$$
$$10y = 0$$
$$y = 0$$

13. $x = 6, y = 0;$
$$(2y + 6) - 3y = 6$$
$$-y + 6 = 6$$
$$y = 0$$

15. $k = {}^{20}\!/_{17}, u = {}^{-6}\!/_{17};$
$$2(8u + 4) + u = 2$$
$$16u + 8 + u = 2$$
$$17u = -6$$
$$u = {}^{-6}\!/_{17}$$

17. $x = 8.1, y = 6.3;$
$$\frac{1}{3}\left(\frac{1}{3}y + 6\right) + y = 9$$
$$\frac{1}{9}y + 2 + y = 9$$
$$\frac{10}{9}y = 7$$
$$y = \frac{63}{10}$$

19. $x = -2, y = -1;$
$$3\left(\frac{1 + 5y}{2}\right) - 2y = -4$$
$$\frac{3}{2} + \frac{15}{2}y - 2y = -4$$
$$\frac{11}{2}y = \frac{-11}{2}$$
$$y = -1$$

21. $r = 4, s = -2;$
$$3\left(\frac{2 - 3s}{2}\right) - 2s = 16$$
$$3 - \frac{9}{2}s - 2s = 16$$
$$\frac{-13}{2}s = 13$$
$$s = -2$$

23. $x = {}^{32}\!/_{31}, k = {}^{-69}\!/_{31};$
$$4\left(4 + \frac{4k}{3}\right) + 5k = -7$$
$$16 + \frac{16k}{3} + 5k = -7$$
$$\frac{31k}{3} = -23$$
$$k = \frac{-69}{31}$$

25. $x = 1000, y = 200;$
$$(800 + y) + y = 1200$$
$$2y = 400$$
$$y = 200$$

27. $r_1 = 1.25, r_2 = 0.75;$
$$(r_2 + 0.50) + r_2 = 2$$
$$2r_2 = 1.5$$
$$r_2 = 0.75$$

29. $V_1 = 40, V_2 = 20;$
$$(2V_2) + V_2 = 60$$
$$3V_2 = 60$$
$$V_2 = 20$$

31. $b = 24, r = 13$;

$$2(50 - 2r) = 3r + 9$$
$$100 - 4r = 3r + 9$$
$$91 = 7r$$
$$r = 13$$

7.3 Addition-Subtraction Method for Solving Simultaneous Equations

1. $x = 5, y = 2$;

$$x + y = 7$$
$$\underline{x - y = 3}$$
$$2x \quad = 10$$

3. $x = 2, y = 1$;

$$2x + y = 5$$
$$\underline{x - y = 1}$$
$$3x \quad = 6$$

5. $m = 4, n = 8$;

$$m + n = 12$$
$$\underline{2m - n = 0}$$
$$3m \quad = 12$$

7. $d = -1, t = 4$;

$$-2d - 2t = -6$$
$$\underline{2d + 3t = 10}$$
$$t = 4$$

9. $x = 1, n = 5$;

$$-3x - 6n = -33$$
$$\underline{3x - 5n = -22}$$
$$-11n = -55$$

11. $x = 9, y = 2$;

$$x - 2y = 5$$
$$\underline{4x + 2y = 40}$$
$$5x \quad = 45$$

13. inconsistent;

$$7x - y = 5$$
$$\underline{-7x + y = -4}$$
$$0 = 1$$

15. $a = 1, b = 2$;

$$-3a - 21b = -45$$
$$\underline{3a + 2b = 7}$$
$$-19b = -38$$

17. $p = \frac{3}{2}, q = \frac{-1}{12}$;

$$26p - 36q = 42$$
$$\underline{12p + 36q = 15}$$
$$38p \quad = 57$$

19. $x = -14, y = -3$;

$$60x - 70y = -630$$
$$\underline{15x + 70y = -420}$$
$$75x \quad = -1050$$

21. $m = -4, n = 6$;

$$6m = 30 - 9n$$
$$\underline{-6m = -24 + 8n}$$
$$0 = 6 - n$$

23. $x = 1, y = 1$;

$$\frac{1}{x} + \frac{2}{y} = 3$$
$$\underline{\frac{1}{x} - \frac{2}{y} = -1}$$
$$\frac{2}{x} \quad = 2$$

25. $V_1 = 6, V_2 = 1.5$;

$$V_1 + V_2 = 7.5$$
$$\underline{V_1 - V_2 = 4.5}$$
$$2V_1 \quad = 12$$

27. $x = \frac{1}{16}, y = 1$;

$$16x + y = 2$$
$$\underline{32x - y = 1}$$
$$48x \quad = 3$$

29. $x = 650$ math score,
     $y = 500$ verbal score;

$$x + y = 1150$$
$$x = y + 150$$

31. $g = 120, s = 25;$

   $30g + 50s = 4850$

   $75g + 25s = 9625$

7.4 Determinants in Two Equations

1. $4(1) - 3(2) = -2$ 3. $2(8) - 3(-5) = 31$ 5. $8(6) - 12(4) = 0$

7. $2(9) - (-5)(-4) = -2$ 9. $-5(-6) - (-8)(-3) = 6$ 11. $10(6) - (-4)(-2) = 52$

13. $x = 1, y = 5;$ $D = \begin{vmatrix} 1 & 1 \\ 1 & -1 \end{vmatrix} = -2,$ $D_x = \begin{vmatrix} 6 & 1 \\ -4 & -1 \end{vmatrix} = -2,$ $D_y = \begin{vmatrix} 1 & 6 \\ 1 & -4 \end{vmatrix} = -10$

15. $x = 2, y = 3;$ $D = \begin{vmatrix} 2 & -1 \\ 1 & 1 \end{vmatrix} = 3,$ $D_x = \begin{vmatrix} 1 & -1 \\ 5 & 1 \end{vmatrix} = 6,$ $D_y = \begin{vmatrix} 2 & 1 \\ 1 & 5 \end{vmatrix} = 9$

17. $s = 4, t = 3;$ $D = \begin{vmatrix} 2 & 0 \\ 1 & -1 \end{vmatrix} = -2,$ $D_s = \begin{vmatrix} 8 & 0 \\ 1 & -1 \end{vmatrix} = -8,$ $D_t = \begin{vmatrix} 2 & 8 \\ 1 & 1 \end{vmatrix} = -6$

19. $V_1 = 5, V_2 = 3;$ $D = \begin{vmatrix} 2 & -1 \\ 3 & -2 \end{vmatrix} = -1,$ $D_{V_1} = \begin{vmatrix} 7 & -1 \\ 9 & -2 \end{vmatrix} = -5,$ $D_{V_2} = \begin{vmatrix} 2 & 7 \\ 3 & 9 \end{vmatrix} = -3$

21. $x = -6, y = 7;$ $D = \begin{vmatrix} 1 & 2 \\ 4 & -5 \end{vmatrix} = -13,$ $D_x = \begin{vmatrix} 8 & 2 \\ -59 & -5 \end{vmatrix} = 78,$ $D_y = \begin{vmatrix} 1 & 8 \\ 4 & -59 \end{vmatrix} = -91$

23. $x = 12, y = 9;$ $D = \begin{vmatrix} 1 & 2 \\ 2 & 1 \end{vmatrix} = -3,$ $D_x = \begin{vmatrix} 30 & 2 \\ 33 & 1 \end{vmatrix} = -36,$ $D_y = \begin{vmatrix} 1 & 30 \\ 2 & 33 \end{vmatrix} = -27$

25. $x = -12, y = -10;$ $D = \begin{vmatrix} 8 & -9 \\ -3 & 4 \end{vmatrix} = 5,$ $D_x = \begin{vmatrix} -6 & -9 \\ -4 & 4 \end{vmatrix} = -60,$ $D_y = \begin{vmatrix} 8 & -6 \\ -3 & -4 \end{vmatrix} = -50$

27. $x = 3, y = \dfrac{1}{2};$ $D = \begin{vmatrix} 3 & 4 \\ 5 & -2 \end{vmatrix} = -26,$ $D_x = \begin{vmatrix} 11 & 4 \\ 14 & -2 \end{vmatrix} = -78,$ $D_y = \begin{vmatrix} 3 & 11 \\ 5 & 14 \end{vmatrix} = -13$

29. $x = 0.2, y = 0.3;$ $D = \begin{vmatrix} 5 & 10 \\ 1 & 1 \end{vmatrix} = -5,$ $D_x = \begin{vmatrix} 4 & 10 \\ 0.5 & 1 \end{vmatrix} = -1,$ $D_y = \begin{vmatrix} 5 & 4 \\ 1 & 0.5 \end{vmatrix} = -1.5$

31. $x = \frac{1}{2}, y = \frac{1}{4};$ $D = \begin{vmatrix} \frac{1}{2} & \frac{1}{2} \\ 1 & -1 \end{vmatrix} = -1, D_x = \begin{vmatrix} \frac{3}{8} & \frac{1}{2} \\ \frac{1}{4} & -1 \end{vmatrix} = -\frac{1}{2},$ $D_y = \begin{vmatrix} \frac{1}{2} & \frac{3}{8} \\ 1 & \frac{1}{4} \end{vmatrix} = -\frac{1}{4}$

33. $R_1 = 24,000, R_2 = 8,000;$ $D = \begin{vmatrix} 1 & 1 \\ 1 & -1 \end{vmatrix} = -2,$ $D_{R_1} = \begin{vmatrix} 32,000 & 1 \\ 16,000 & -1 \end{vmatrix} = -48,000,$ $D_{R_2} = \begin{vmatrix} 1 & 32,000 \\ 1 & 16,000 \end{vmatrix} = -16,000$

35. $x = \frac{16}{3}, y = \frac{32}{3};$ $D = \begin{vmatrix} 0.3 & 0.6 \\ 0.7 & 0.4 \end{vmatrix} = -0.30,$ $D_x = \begin{vmatrix} 8 & 0.6 \\ 8 & 0.4 \end{vmatrix} = -1.6,$ $D_y = \begin{vmatrix} 0.3 & 8 \\ 0.7 & 8 \end{vmatrix} = -3.2$

37. 4.50 for 1 file, 0.75 for 1 blade;

$6x + 3y = 29.25$

$2x + 4y = 12.00$

39. 40.2 ohms and 15.5 ohms;

$R_1 + R_2 = 55.7$

$R_1 - R_2 = 24.7$

41. $x = 1.428, y = -0.653; D = \begin{vmatrix} 2.56 & -3.47 \\ 3.76 & 1.93 \end{vmatrix} = 17.988, D_x = \begin{vmatrix} 5.92 & -3.47 \\ 4.11 & 1.93 \end{vmatrix} = 25.6873, D_y = \begin{vmatrix} 2.56 & 5.92 \\ 3.76 & 4.11 \end{vmatrix} = -11.7376$

43. $x = 0.00609, y = 0.00150;$ $D = \begin{vmatrix} 3725 & -4290 \\ 4193 & 2558 \end{vmatrix} = 27,516,520,$ $D_x = \begin{vmatrix} 16.25 & -4290 \\ 29.36 & 2558 \end{vmatrix} = 167,521.9,$

$D_y = \begin{vmatrix} 3725 & 16.25 \\ 4193 & 29.36 \end{vmatrix} = 41,229.75$

7.5 Problem Solving Using Systems of Linear Equations

1. $A_1 = 8A, A_2 = -3A;$

$A_1 + A_2 = 5$

$A_1 - A_2 = 11$

3. $x = 30$ hours, $y = 75$ hours;

$5x = 2y$

$x + y = 105$

5. $x = 8$ mm, $y = 21$ mm;

$y = 2x + 5$

$4y = 4x + 52$

7. $x = \$8000, y = \$200$

$x + y = 8200$

$0.04x + 0.05y = 330$

9. $x = 90,000$ BTU at 80%,

$y = 60,000$ BTU at 70%;

$x + y = 150,000$

$0.80x + 0.70y = 114,000$

11. $x = 42\%, y = 58\%$
$$x + y = 100$$
$$x = y - 16$$

13. $x = 48$ teeth, $y = 16$ teeth;
$$x + y = 64$$
$$x = 3y$$

15. $x = 80$ g, $y = 40$ g;
$$0.60x + 0.30y = 60$$
$$0.40x + 0.70y = 60$$

17. $x = 210$ cell phones,
$y = 110$ radar detectors;
$$x + y = 320$$
$$110x + 160y = 40,700$$

19. $x = 100$ breadboards,
$y = 60$ testers;
$$7x + 5y = 1000$$
$$4x + 2.5y = 550$$

Review Exercises for Chapter 7

1. $x = \dfrac{12}{5}, y = \dfrac{6}{5}$;
$$2x + y = 6$$
$$\underline{3x - y = 6}$$
$$5x \quad = 12$$

3. $x = 4, y = 4$;
$$-x - 2y = -12$$
$$\underline{x + 3y = 16}$$
$$y = 4$$

5. $x = 4, y = 4$;
$$6x - 3y = 12$$
$$\underline{x + 3y = 16}$$
$$7x \quad = 28$$

7. $x = \dfrac{18}{5}, y = \dfrac{2}{5}$; $D = \begin{vmatrix} 2 & 7 \\ 3 & -2 \end{vmatrix} = -25,$ $D_x = \begin{vmatrix} 10 & 7 \\ 10 & -2 \end{vmatrix} = -90,$ $D_y = \begin{vmatrix} 2 & 10 \\ 3 & 10 \end{vmatrix} = -10$

9. $u = 1, v = -1$;
$$2u - 3v = 5$$
$$\underline{-2u + 8v = -10}$$
$$5v = -5$$

11. $a = 12, b = -3$;
$$6a + 14b = 30$$
$$\underline{-6a + 15b = -117}$$
$$29b = -87$$

13. $x = -9, y = 11$;
$$7x + 5(-3x - 16) = -8$$
$$7x - 15x - 80 = -8$$
$$-8x = 72$$

15. $y = 1, z = -\dfrac{1}{2}$;
$$-30y + 20z = -40$$
$$\underline{30y + 36z = 12}$$
$$56z = -28$$

17.   dependent;
$$-6n - 2d = -40$$
$$\underline{6n + 2d = 40}$$
$$0 = 0$$

19. $x = -\dfrac{7}{3}, y = 2$;
$$-6x - 2y = 10$$
$$\underline{6x + 8y = 2}$$
$$6y = 12$$

21. $x = \dfrac{43}{19}; y = -\dfrac{22}{19}$; $D = \begin{vmatrix} 2 & -3 \\ 5 & 2 \end{vmatrix} = 19,$ $D_x = \begin{vmatrix} 8 & -3 \\ 9 & 2 \end{vmatrix} = 43,$ $D_y = \begin{vmatrix} 2 & 8 \\ 5 & 9 \end{vmatrix} = -22$

23. $x = \dfrac{94}{107}, y = -\dfrac{22}{107}$; $D = \begin{vmatrix} 8 & 5 \\ 7 & -9 \end{vmatrix} = -107,$ $D_x = \begin{vmatrix} 6 & 5 \\ 8 & -9 \end{vmatrix} = -94,$ $D_y = \begin{vmatrix} 8 & 6 \\ 7 & 8 \end{vmatrix} = 22$

25. $x = 100, \ y = -1$;

$$0.09x + 6y = 3$$
$$\underline{0.04x - 6y = 10}$$
$$0.13x \qquad = 13$$

27. $x = \dfrac{119}{201}, \ y = \dfrac{-59}{201}$; $D = \begin{vmatrix} 2 & -13 \\ 15 & 3 \end{vmatrix} = 201,$ $\qquad D_x = \begin{vmatrix} 5 & -13 \\ 8 & 3 \end{vmatrix} = 119,$ $\qquad D_y = \begin{vmatrix} 2 & 5 \\ 15 & 8 \end{vmatrix} = -59$

29. $x = 12, \ y = 24$;

$$\frac{2}{3}x + \frac{3}{4}y = 26$$
$$\underline{\frac{1}{3}x - \frac{3}{4}y = -14}$$
$$x \qquad\quad = 12$$

31. $r = \dfrac{1}{3}, \ s = \dfrac{1}{5}$;

$$\frac{27}{r} - \frac{15}{s} = 6$$
$$\underline{\frac{-35}{r} + \frac{15}{s} = -30}$$
$$\frac{-8}{r} \qquad = -24$$

33. $x = 1, \ y = 0.3$

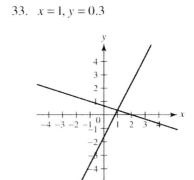

35. $x = 1.6, \ y = -1.2$

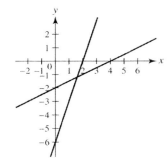

37. $u = 2, \ v = -6$

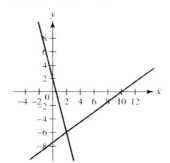

39. parallel lines, inconsistent system

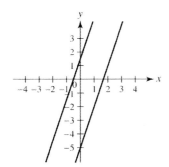

41. $10 - 12 = -2$

43. $-15 + 8 = -7$

45. $x = -7, y = 2;$ $D = \begin{vmatrix} 1 & 3 \\ 2 & 8 \end{vmatrix} = 2,$ $D_x = \begin{vmatrix} -1 & 3 \\ 2 & 8 \end{vmatrix} = -14,$ $D_y = \begin{vmatrix} 1 & -1 \\ 2 & 2 \end{vmatrix} = 4$

47. $x = \frac{1}{2}, y = 4;$ $D = \begin{vmatrix} 2 & 3 \\ 6 & -2 \end{vmatrix} = -22,$ $D_x = \begin{vmatrix} 13 & 3 \\ -5 & -2 \end{vmatrix} = -11,$ $D_y = \begin{vmatrix} 2 & 13 \\ 6 & -5 \end{vmatrix} = -88$

49. $i_1 = -\frac{3}{22}, i_2 = -\frac{39}{110};$

$$12i_1 - 30i_2 = 9$$
$$10i_1 + 30i_2 = -12$$
$$\overline{22i_1 \qquad = -3}$$

51. $m = 416.73, n = 330.97;$ $D = \begin{vmatrix} 1.58 & -2 \\ 0.424 & -0.0728 \end{vmatrix} = 0.732976,$ $D_m = \begin{vmatrix} -3.5 & -2 \\ 152.6 & -0.0728 \end{vmatrix} = 305.4548,$

$$D_n = \begin{vmatrix} 1.58 & -3.5 \\ 0.424 & 152.6 \end{vmatrix} = 242.592$$

53. $x = 45$ spots, $y = 30$ spots;
$$3x + 4y = 255$$
$$4x + 3y = 270$$

55. $x = 112$ require additional payments,
$y = 72$ require refunds;
$$x + y = 184$$
$$x = 2y - 32$$

57. $x = 45$ mph, $y = 55$ mph;
$$3x + 4y = 355$$
$$4x + 3y = 345$$

59. $x = 750$ mL of 5%,
$y = 250$ mL of 25%;
$$x + y = 1000$$
$$0.05x + 0.25y = 0.10(1000)$$

Chapter 7 Test

1. $x = 2.2, y = 1.2$

2. $x = 7, y = -2;$
$$-2x - 2y = -10$$
$$\underline{2x + 4y = 6}$$
$$2y = -4$$

3. $x = 2, y = -4;$
$$7x - 2(4 - 4x) = 22$$
$$7x - 8 + 8x = 22$$
$$15x = 30$$

4. $14 + 5 = 19$

5. $x = 2, y = 1;$ $D = \begin{vmatrix} 3 & -2 \\ 2 & 5 \end{vmatrix} = 19,$ $D_x = \begin{vmatrix} 4 & -2 \\ 9 & 5 \end{vmatrix} = 38,$ $D_y = \begin{vmatrix} 3 & 4 \\ 2 & 9 \end{vmatrix} = 19$

6. $x = {5650}/{2401}, y = {14,960}/{2401};$ $D = \begin{vmatrix} 22 & 27 \\ -27 & 76 \end{vmatrix} = 2401,$ $D_x = \begin{vmatrix} 220 & 27 \\ 410 & 76 \end{vmatrix} = 5650,$

$D_y = \begin{vmatrix} 22 & 220 \\ -27 & 410 \end{vmatrix} = 14,960$

7. $x = \$58$ for 2-inch sets,
   $y = \$75$ for ${5}/{4}$-inch sets;
   $6x + 13y = 1323$
   $3x + 2y = 324$

Chapter 8
Factoring

8.1 The Distribution Property and Common Factors

1. $5(x + y)$

3. $7(a^2 - 2bc)$

5. $a(a + 2)$

7. $2x(x - 2)$

9. $3(ab - c)$

11. $2p(2 - 3q)$

13. $3y^2(1 - 3z)$

15. $abx(1 - xy)$

17. $6x(1 - 3y)$

19. $3ab(a + 3)$

21. $acf(abc - 4)$

23. $ax^2y^2(x + y)$

25. $2(x + y - z)$

27. $5(x^2 + 3xy - 4y^3)$

29. $2x(3x + 2y - 4)$

31. $4pq(3q - 2 - 7q^2)$

33. $7a^2b^2(5ab^2c^2 + 2b^3c^3 - 3a)$

35. $3a(2ab - 1 + 3b^2 - 4ab^2)$

37. $(x^3 + 3x^2) + (2x + 6)$
$x^2(x + 3) + 2(x + 3)$
$(x + 3)(x^2 + 2)$

39. $(3x^3 - 15x^2) + (5x - 25)$
$3x^2(x - 5) + 5(x - 5)$
$(x - 5)(3x^2 + 5)$

41. $(xy + x) + (3y + 3)$
$x(y + 1) + 3(y + 1)$
$(y + 1)(x + 3)$

43. $\dfrac{nRT_2}{T_1} - nR$
$nR\left(\dfrac{T_2}{T_1} - 1\right)$

45. $V = 6wh + 8wh + 4w^2h$
$V = 2wh(3 + 4 + 2w)$
$V = 2wh(7 + 2w)$

47. $P = 2x^3 - 6x^2 + 10x$
$P = 2x(x^2 - 3x + 5)$

49. a) yes
b) no; $2x(x - 4)$

51. a) yes
b) no; $12ay(xy - 3)$

53. a) prime
b) no; $4(x - 2)$
c) no, $3(x - 3)$

55. a) no; $y(x + 1)$
b) prime
c) no, $3y(x + 3)$

8.2 Factoring Trinomials

1. $p = 1(2) = 2$
   $s = 3$
   $2 \times 1 = 2; 2 + 1 = 3$
   $(x + 2)(x + 1)$

3. $p = 1(-12) = -12$
   $s = 1$
   $4 \times -3 = -12; 4 + -3 = 1$
   $(x - 3)(x + 4)$

5. $p = 1(-5) = -5$
   $s = -4$
   $-5 \times 1 = -5; -5 + 1 = -4$
   $(y - 5)(y + 1)$

7. $p = 1(25) = 25$
   $s = 10$
   $5 \times 5 = 25; 5 + 5 = 10$
   $(x + 5)(x + 5)$

9. $p = 1(14) = 14$
   $s = 9$
   $7 \times 2 = 14; 2 + 7 = 9$
   $(x + 2)(x + 7)$

11. $p = 1(-42) = -42$
    $s = -1$
    $-7 \times 6 = -42; -7 + 6 = -1$
    $(x + 6)(x - 7)$

13. $p = 1(32) = 32$
    $s = 12$
    $4 \times 8 = 32; 4 + 8 = 12$
    $(x + 8)(x + 4)$

15. $p = 1(1) = 1$
    $s = 2$
    $1 \times 1 = 1; 1 + 1 = 2$
    $(x + 1)(x + 1)$

17. $p = 1(15) = 15$
    $s = -8$
    $-5 \times -3 = 15; -5 + -3 = -8$
    $(x - 5)(x - 3)$

19. $p = 1(-14) = -14$
    $s = 5$
    $7 \times -2 = -14; 7 + -2 = 5$
    $(x - 2)(x + 7)$

21. $p = 1(12) = 12$
    $s = 8$
    $2 \times 6 = 12; 2 + 6 = 8$
    $(x + 6)(x + 2)$

23. $p = 1(-40) = -40$
    $s = -6$
    $4 \times -10 = -40; 4 + -10 = -6$
    $(x + 4)(x - 10)$

25. $p = 1(-108) = -108$
    $s = -3$
    $9 \times -12 = -108; 9 + -12 = -3$
    $(x - 12)(x + 9)$

27. $(x + 1)(x + 1) = (x + 1)^2$

29. $(x - 4)(x - 4) = (x - 4)^2$

31. $N(r^2 + 2r + 1)$
    $N(r + 1)(r + 1)$

33. $P\left(R^2 + 2RH\right) = P\left(R + 1\right)\left(R + 1\right) = P\left(R + 1\right)^2$

8.3 Factoring General Trinomials

1. $(2q + 1)(q + 5)$

3. $(3x + 1)(x - 3)$

5. $(5c - 1)(c + 7)$

7. prime

9. $(2s - 3t)(s - 5t)$

11. $(5x + 2)(x + 3)$

13. $(2x - 3)(2x - 1)$

15. $(2q + 3)(6q + 1)$

17. $(6t - 5u)(t + 2u)$

19. $(2x + 3)(4x - 3)$

21. $(4x - 3)(x + 6)$

23. $(4n - 5)(2n + 3)$

25. $2(x^2 - 11x + 24)$
$2(x - 8)(x - 3)$

27. $2(2x^2 + xz - 6z^2)$
$2(2x - 3z)(x + 2z)$

29. $2x(x^2 + 3x + 2)$
$2x(x + 1)(x + 2)$

31. $a(10x^2 + 23xy - 5y^2)$
$a(5x - y)(2x + 5y)$

33. $3a(x^2 + 2x - 15)$
$3a(x + 5)(x - 3)$

35. $7a^3(2x^2 - x - 1)$
$7a^3(2x + 1)(x - 1)$

37. $(2x + 1)(2x + 1)$

39. $(3x - 1)(3x - 1)$

41. $2(p^2 - 54p + 200)$
$2(p - 50)(p - 4)$

43. $2(x^2 - 12x + 32)$
$2(x - 4)(x - 8)$

45. $4x(x^2 - 12x + 36)$
$4x(x - 6)(x - 6)$

47. $3x(2x^2 - 5x + 2)$
$3x(2x - 1)(x - 2)$

49. $4(4t^2 - 31t - 8)$
$4(4t + 1)(t - 8)$

8.4 The Difference Between Two Squares

1. $(a - 1)(a + 1)$

3. $(t - 3)(t + 3)$

5. $(4 - x)(4 + x)$

7. $\left(2x^2\right)^2 - y^2$
$\left(2x^2 - y\right)\left(2x^2 + y\right)$

9. $(10)^2 - \left(a^2b\right)^2$
$\left(10 - a^2b\right)\left(10 + a^2b\right)$

11. $(ab)^2 - (y)^2$
$(ab - y)(ab + y)$

13. $\left(9x^2\right)^2 - \left(2y^3\right)^2$

    $\left(9x^2 - 2y^3\right)\left(9x^2 + 2y^3\right)$

15. $5(x^4 - 9)$

    $5\left(x^2\right)^2 - (3)^2$

    $5\left(x^2 - 3\right)\left(x^2 + 3\right)$

17. $4\left[(x)^2 - (5y)^2\right]$

    $4(x - 5y)(x + 5y)$

19. $\left(x^2\right)^2 - (1)^2$

    $\left(x^2 - 1\right)\left(x^2 + 1\right)$

    $(x - 1)(x + 1)(x^2 + 1)$

21. $4(x + 9y^2)$

23. $3s^2(t^4 + 4)$

25. $\left[5 - (x + y)\right]\left[5 + (x + y)\right]$

    $(5 - x - y)(5 + x + y)$

27. $\left[(x + y) - (x - y)\right]\left[(x + y) + (x - y)\right]$

    $(2y)(2x)$

    $4xy$

29. $a\left[(x + y)^2 - 1^2\right]$

    $a(x + y + 1)(x + y - 1)$

8.5 The Sum and Differences of Cubes

1. $(a - 1)\left(a^2 + a + 1\right)$

3. $(t + 2)\left(t^2 - 2t + 4\right)$

5. $(1 - x)\left(1 + x + x^2\right)$

7. $(2x)^3 + (3a)^3$

    $(2x + 3a)(2^2x^2 - (2x)(3a) + 3^2a^2)$

    $(2x + 3a)(4x^2 - 6ax + 9a^2)$

9. $\left(2x^2\right)^3 - (y)^3$

    $(2x^2 - y)(2^2x^4 + \left(2x^2\right)(y) + y^2)$

    $(2x^2 - y)(4x^4 + 2x^2y + y^2)$

11. $(ax)^3 - \left(y^2\right)^3$

    $(ax - y^2)(a^2x^2 + axy^2 + (y^2)^2)$

    $(ax - y^2)(a^2x^2 + axy^2 + y^4)$

13. $8x(x^3 + 1)$

    $8x(x + 1)(x^2 - x + 1)$

15. $ax^2(1 - y^3)$

    $ax^2(1 - y)(1 + y + y^2)$

17. $2k(R_1^3 + 8R_2^3)$

    $2k\left[(R_1)^3 + (2R_2)^3\right]$

    $2k(R_1 + 2R_2)(R_1^2 - 2R_1R_2 + 4R_2^2)$

19. $2(27t_1^6 - t_2^3)$

    $2\left[\left(3t_1^2\right)^3 - (t_2)^3\right]$

    $2(3t_1^2 - t_2)(9t_1^4 + 3t_1^2t_2 + t_2^2)$

21. $\left(5s^2\right)^3 - \left(4t^3\right)^3$

$(5s^2 - 4t^3)(25s^4 + 20s^2t^3 + 16t^6)$

23. $(ab)^3 + \left(c^5\right)^3$

$(ab + c^5)(a^2b^2 - abc^5 + c^{10})$

25. $(1)^2 - \left(a^3x^3\right)^2$

$(1 - a^3x^3)(1 + a^3x^3)$

$(1 - ax)(1 + ax + a^2x^2)(1 + ax)(1 - ax + a^2x^2)$

27. $\left[1 - (x + y)\right]\left[1 + (x + y) + (x + y)^2\right]$

$\left[1 - x - y\right]\left[1 + x + y + x^2 + 2xy + y^2\right]$

29. $N\left(x^3 - y^3\right)$

$N(x - y)\left(x^2 + xy + y^2\right)$

31. $(2x + 3)\left(4x^2 - 6x + 9\right)$

33. $at^3\left(t^3 - 1\right)$

$at^3(t - 1)\left(t^2 + t + 1\right)$

Review Exercises

1. $5(a - c)$

3. $3a(a + 2)$

5. $4ab(3a + 1)$

7. $8stu^2(1 - 3s^2)$

9. $(2x - y)(2x + y)$

11. $\left(4y^2\right)^2 - \left(x\right)^2$

$(4y^2 - x)\left(4y^2 + x\right)$

13. $(x + 1)(x + 1)$

15. $(x - 6)(x - 1)$

17. $a\left(x^2 + 3ax - a^2\right)$

19. $2nm\left(m^2 - 2mn + 3n^2\right)$

21. $4t^2\left(p^3 - 3t^2 - 1 + a\right)$

23. $2xy^3\left(1 - 7x + 8y - 3x^2y^2\right)$

25. $\left(4rs - 3y\right)\left(4rs + 3y\right)$

27. prime

29. $(2x + 7)(x + 1)$

31. $(5s + 2)(s - 1)$

33. $(2t - 3)(7t + 1)$

35. $(3x + 1)(3x + 1)$

37. $(x + 2y)(x + y)$

39. $(5c - d)(2c + 5d)$

41. $(8x + 7)(11x - 12)$

43. $2\left(x^2 - 9y^2\right)$

$2(x + 3y)\left(x - 3y\right)$

45. $8y^2x^4\left[(xy)^2-(2)^2\right]$
$8y^2x^4(xy+2)(xy-2)$

47. $3a\left(x^2+x-12\right)$
$3a(x+4)(x-3)$

49. $3r\left(27r^2-15rs-52s^2\right)$
$3r(6r-13s)(3r+4s)$

51. $16y^3\left(3-4y+y^2\right)$
$16y^3(y-1)(y-3)$

53. $5\left[\left(x^2\right)^2-(5)^2\right]$
$5\left(x^2-5\right)\left(x^2+5\right)$

55. $\left(4x^2\right)^2-(1)^2$
$\left(4x^2-1\right)\left(4x^2+1\right)$
$(2x-1)(2x+1)\left(4x^2+1\right)$

57. x^3+3^3
$(x+3)\left(x^2-3x+9\right)$

59. $(2x)^3+1^3$
$(2x+1)\left(4x^2-2x+1\right)$

61. $axy\left(x^3-y^3\right)$
$axy(x-y)\left(x^2+xy+y^2\right)$

63. $i\left(R_1+R_2+R_3\right)$

65. $P(N+2)$

67. $k\left(D^2-(2r)^2\right)$
$k(D+2r)(D-2r)$

69. $b\left[\left(x^2+y^2\right)-2y^2\right]$

Chapter Test

1. $6(x-4y)$

2. $3xy(4+y+2xy)$

3. $\overbrace{xy-x}+\overbrace{3y-3}$
$x(y-1)+3(y-1)$
$(y-1)(x+3)$

4. $(x+8)(x+3)$

5. $(7x-2)(x-1)$

6. $(2x-5)(2x+5)$

7. $8\left(x^3-8y^3\right)$
$8(x-2y)\left(x^2+2xy+4y^2\right)$

8. $x\left(x^3+3^3\right)$
$x(x+3)\left(x^2-3x+9\right)$

9. $(R-300)(R-100)$

10. $16\left(x^2-7x+6\right)$
$16(x-6)(x-1)$

Chapter 9
Algebraic Fractions

9.1 Equivalent Fractions

1. $\dfrac{4}{7}$

3. $\dfrac{2a}{3a^2}$

5. $\dfrac{2}{x+1}$

7. $\dfrac{(2x-1)(x+3)}{(x+3)(x+1)} = \dfrac{2x-1}{x+1}$

9. $\dfrac{3(x^2-4)}{(x+2)(x+2)} = \dfrac{3(x+2)(x-2)}{(x+2)(x+2)} = \dfrac{3(x-2)}{x+2}$

11. $\dfrac{(2-x)(2+x)}{(x-3)(x-2)} = \dfrac{-(-2+x)(2+x)}{(x-3)(x-2)} = \dfrac{-(2+x)}{(x-3)}$

13. $\dfrac{1}{3}$

15. $\dfrac{ab}{4}$

17. $\dfrac{8}{9}$

19. $\dfrac{2x-1}{x-2}$

21. $\dfrac{(x+1)(x+2)}{2(x+3)}$

23. $\dfrac{(x-1)(x+1)}{(x-1)(x-1)} = \dfrac{(x+1)}{x-1}$

25. $\dfrac{x(3x-1)}{(x+2)(3x-1)} = \dfrac{x}{x+2}$

27. $\dfrac{(2x-5)(3x-2)}{(2x-5)(4x+3)} = \dfrac{3x-2}{4x+3}$

29. $\dfrac{(x-3y)(x+3y)}{3y(x-3y)} = \dfrac{x+3y}{3y}$

31. $\dfrac{(4-x)(5-x)}{(4-x)(2+x)} = \dfrac{5-x}{2+x}$

33. $\dfrac{-3x(-1+3x)}{3x-1} = -3x$

35. $\dfrac{(x-2)(x-3)}{-(-3+x)(3+x)} = \dfrac{x-2}{-(3+x)}$

37. $\dfrac{4(a-2)}{4(a+2)} = \dfrac{a-2}{a+2} = \dfrac{5-2}{5+2} = \dfrac{3}{7}$

$\dfrac{4(5)-8}{4(5+2)} = \dfrac{20-8}{4(7)} = \dfrac{12}{28} = \dfrac{3}{7}$

9.2 Equivalent Fractions

1. $\dfrac{1}{8n}; 13s$

3. $\dfrac{3b}{a}; \dfrac{a}{3b}$

5. $\dfrac{x-y}{x+y}; \dfrac{x^2}{x^2+y^2}$

7. $\dfrac{a+b}{a}; \dfrac{-V}{IR}$

9. $\dfrac{\overset{4}{\cancel{12}}}{\underset{3}{\cancel{18}}} \bullet \dfrac{\overset{1}{\cancel{6}}}{\underset{3}{\cancel{9}}t} = \dfrac{4}{9t}$

11. $\dfrac{2}{\underset{3}{\cancel{9}}} \bullet \dfrac{\overset{1}{\cancel{3}}a}{5} = \dfrac{2a}{15}$

13. $\dfrac{1\cancel{7}r\cancel{s}}{\underset{4}{1\cancel{2}\cancel{t}}} \bullet \dfrac{\cancel{3}t^{\cancel{2}1}}{\underset{3}{5\cancel{t}\cancel{s}}} = \dfrac{rt}{12}$

15. $\left(\dfrac{3}{4}\right)^4 = \dfrac{3^4}{4^4} = \dfrac{81}{256}$

17. $\left(\dfrac{a^2}{2x}\right)^5 = \dfrac{a^{10}}{2^5 x^5} = \dfrac{a^{10}}{32x^5}$

19. $\dfrac{a^3 x^9}{b^6}$

21. $\dfrac{2}{5x} \bullet \dfrac{13}{7c} = \dfrac{26}{35xc}$

23. $\dfrac{6x}{1\cancel{7}} \bullet \dfrac{6\cancel{8}^4 m}{7} = \dfrac{24xm}{7}$

25. $\dfrac{\cancel{3}x}{2\cancel{5}y} \bullet \dfrac{\cancel{5}y^{\cancel{2}1}}{2\cancel{7}x^{\cancel{2}1}} = \dfrac{y}{45x}$ $\;\underset{5}{}\;\underset{9}{}$

27. $\dfrac{\cancel{9}a^2 \cancel{b}^2}{1\cancel{6}ab^{\cancel{3}1}} \bullet \dfrac{4\cancel{0}^4}{7\cancel{2}a^{\cancel{3}2}b^4} = \dfrac{4}{8a^2 b^5} = \dfrac{1}{2a^2 b^5}$ $\;\underset{8}{}$

29. $\dfrac{a-5b}{a+b} \bullet \dfrac{a+3b}{a-5b} = \dfrac{a+3b}{a+b}$

31. $\dfrac{(x-y)(x\cancel{+}y)}{\underset{2}{1\cancel{4}\cancel{x}}} \bullet \dfrac{3\cancel{5}x^{\cancel{2}1}}{3(x\cancel{+}y)} = \dfrac{5x(x-y)}{6}$

33. $\dfrac{\cancel{x}}{(x\cancel{-}1)} \bullet \dfrac{(x\cancel{-}1)(x+1)}{\cancel{x}(x+2)} = \dfrac{x+1}{x+2}$

35. $\dfrac{(x+3)(x\cancel{-}1)}{(x-2)(x\cancel{+}2)} \bullet \dfrac{(x-3)(x\cancel{+}2)}{(x-4)(x\cancel{-}1)} = \dfrac{(x+3)(x-3)}{(x-2)(x-4)}$

37. $\dfrac{2b\cancel{+}3}{5} \bullet \dfrac{5b-2}{2(2b\cancel{+}3)} = \dfrac{5b-2}{10}$

39. $\dfrac{3(a^2-b^2)}{(a-2b)(a+2b)} \bullet \dfrac{a+2b}{(a+b)(a+b)}$

$\dfrac{3(a-b)(a\cancel{+}b)}{(a-2b)(a\cancel{+}2b)} \bullet \dfrac{a\cancel{+}2b}{(a\cancel{+}b)(a+b)}$

$\dfrac{3(a-b)}{(a-2b)(a+b)}$

41. $\dfrac{(s - 7)(s+2)}{(s-12)(s \cancel{+} 3)} \cdot \dfrac{(s+7)(s \cancel{+} 3)}{(s - 7)(s+11)} = \dfrac{(s+2)(s+7)}{(s-12)(s+11)}$

43. $\dfrac{3.2 \cancel{x} (x+2)}{x^{\cancel{2}1}} = \dfrac{3.2(x+2)}{x}$

45. $\dfrac{(1-n)(1-n)}{(1+n)(1+n)} = \dfrac{1^{\cancel{2}} - 2n + n^2}{1 + 2n + n^2}$

9.3 The Lowest Common Denominator

1. 18

3. 36

5. 12a

7. 40t

9. 90y

11. $9x^2$

13. $8x^2$

15. 420ax

17. $375ax^2$

19. $75a^3$

21. $96a^3b^3$

23. $15a^2$

25. $60a^2cx^3$

27. $\dfrac{5}{4(x-1)} ; \dfrac{3}{8x}$

   $LCD = 8x(x-1)$

29. $\dfrac{4}{3(a+3)} ; \dfrac{5}{a(a+3)}$

   $LCD = 3a(a+3)$

31. $\dfrac{3x}{2(x-y)} ; \dfrac{5}{x(x-y)} ; \dfrac{7x}{6(x-y)(x+y)}$

   $LCD = 6x(x-y)(x+y)$

33. $\dfrac{x-5}{(x-2)(x-1)} ; \dfrac{x}{2(x^2 - 2x + 1)}$

   $\dfrac{x-5}{(x-2)(x-1)} ; \dfrac{x}{2(x-1)(x-1)}$

   $LCD = 2(x-1)^2 (x-2)$

35. $\dfrac{7}{(2t+3)(t-4)} ; \dfrac{5t}{2(t^2 + 5t + 3)}$

   $LCD = 2(2t+3)(t-4)(t^2 + 5t + 3)$

37. $\dfrac{x}{(x-3)(x+2)} ; \dfrac{2x}{(x+3)(x+3)} ; \dfrac{x}{(x-3)(x+3)}$

   $LCD = (x+3)^2 (x-3)(x+2)$

39. $\dfrac{5}{x\left(2x^2-3x+1\right)};\dfrac{x}{x^2\left(x^2-1\right)};\dfrac{2-x}{\left(2x-1\right)\left(x+1\right)}$

$\text{LCD}=x^2\left(2x-1\right)\left(x-1\right)\left(x+1\right)$

9.4 Addition and Subtraction of Algebraic Fractions

1. $\dfrac{5(4)}{36a}-\dfrac{7(3)}{36a}=\dfrac{20}{36a}-\dfrac{21}{36a}$

3. $\dfrac{5b}{abx}+\dfrac{a}{abx}-\dfrac{4bx}{abx}$

5. $\dfrac{4}{x(x-1)}-\dfrac{3}{2x^2(x-1)}$

$\dfrac{4(2x)}{2x^2(x-1)}-\dfrac{3}{2x^2(x-1)}=\dfrac{8x}{2x^2(x-1)}-\dfrac{3}{2x^2(x-1)}$

7. $\dfrac{x}{2(x-2)}+\dfrac{5}{(x-2)(x+2)}-\dfrac{3x}{(x+2)(x+2)}$

$\dfrac{x(x+2)^2}{2(x+2)^2(x-2)}+\dfrac{5(2)(x+2)}{2(x+2)^2(x-2)}-\dfrac{3x(2)(x-2)}{2(x+2)^2(x-2)}$

9. $\dfrac{2(2)}{10x}+\dfrac{3(5)}{10x}=\dfrac{4+15}{10x}=\dfrac{19}{10x}$

11. $\dfrac{2b}{3ab}-\dfrac{5a}{3ab}=\dfrac{2b-5a}{3ab}$

13. $\dfrac{6x}{x^2}+\dfrac{3}{x^2}=\dfrac{6x+3}{x^2}$

15. $\dfrac{20b}{40b}-\dfrac{25}{40b}+\dfrac{6b}{40b}=\dfrac{20b-25+6b}{40b}=\dfrac{26b-25}{40b}$

17. $\dfrac{8y}{by^2}-\dfrac{b}{by^2}=\dfrac{8y-b}{by^2}$

19. $\dfrac{3(3)}{3x^3y}+\dfrac{2x}{3x^3y}=\dfrac{9+2x}{3x^3y}$

21. $\dfrac{2xy}{x^2y}+\dfrac{5x^2}{x^2y}-\dfrac{3}{x^2y}=\dfrac{2xy+5x^2-3}{x^2y}$

23. $\dfrac{7(6yz)}{12xyz}-\dfrac{5(3xz)}{12xyz}+\dfrac{2xy}{12xyz}=\dfrac{42yz-15xz+2xy}{12xyz}$

25. $\dfrac{3(a+2)}{(a-2)(a+2)} + \dfrac{5(a-2)}{(a-2)(a+2)}$

$\dfrac{3a+6+5a-10}{(a-2)(a+2)}$

$\dfrac{8a-4}{(a-2)(a+2)}$

$\dfrac{4(2a-1)}{(a-2)(a+2)}$

27. $\dfrac{x}{2(x-3)} - \dfrac{3}{4(x+3)}$

$\dfrac{2x(x+3)}{4(x-3)(x+3)} - \dfrac{3(x-3)}{4(x-3)(x+3)}$

$\dfrac{2x^2+6x-3x+9}{4(x-3)(x+3)}$

$\dfrac{2x^2+3x+9}{4(x-3)(x+3)}$

29. $\dfrac{4}{(2-3x)(2+3x)} - \dfrac{x-5}{(2+3x)}$

$\dfrac{4}{(2-3x)(2+3x)} - \dfrac{(x-5)(2-3x)}{(2+3x)(2-3x)}$

$\dfrac{4-2x+3x^2+10-15x}{(2+3x)(2-3x)}$

$\dfrac{3x^2-17x+14}{(2+3x)(2-3x)}$

31. $\dfrac{3x}{(x-2)(x+2)} + \dfrac{2}{(x-2)} - \dfrac{5x}{x+2}$

$\dfrac{3x}{(x-2)(x+2)} + \dfrac{2(x+2)}{(x-2)(x+2)} - \dfrac{5x(x-2)}{(x-2)(x+2)}$

$\dfrac{3x+2x+4-5x^2+10x}{(x-2)(x+2)}$

$\dfrac{-5x^2+15x+4}{(x-2)(x+2)}$

33. $\dfrac{2(x+5)(x-5)}{3(x+5)(x-5)} + \dfrac{3(3)(x-5)}{3(x+5)(x-5)} - \dfrac{2(3)(x+5)}{3(x+5)(x-5)}$

$\dfrac{2x^2-50+9x-45-6x-30}{3(x+5)(x-5)}$

$\dfrac{2x^2+3x-125}{3(x+5)(x-5)}$

35. $26.500 + 5.080x + \dfrac{0.004}{x^2}$

$\dfrac{26.500x^2}{x^2} + \dfrac{5.080x^3}{x^2} + \dfrac{0.004}{x^2}$

$\dfrac{26.500x^2+5.080x^3+0.004}{x^2}$

37. $\dfrac{p^2}{2mr^2} - \dfrac{gmM}{r}$

$\dfrac{p^2}{2mr^2} - \dfrac{2gm^2Mr}{2mr^2} = \dfrac{p^2-2gm^2Mr}{2mr^2}$

39. $\dfrac{R_2}{R_1R_2} + \dfrac{R_1}{R_1R_2} = \dfrac{R_2+R_1}{R_1R_2}$

41. $\dfrac{u}{uu_d} - \dfrac{u_d}{uu_d} = \dfrac{u-u_d}{uu_d}$

9.5 Solving Fractional Equations

1. $2x = 8$
   $x = 4$

3. $4\left(\dfrac{2s}{4} - 1\right) = 4s$
   $2s - 4 = 4s$
   $-4 = 2s$
   $s = -2$

5. $10\left(\dfrac{2x}{5} + 3\right) = \dfrac{7x}{10}(10)$
   $4x + 30 = 7x$
   $-3x = -30$
   $x = 10$

7. $8\left(\dfrac{t+1}{2} + \dfrac{1}{4}\right) = \dfrac{3}{8}(8)$
   $4(t+1) + 2 = 3$
   $4t + 4 + 2 = 3$
   $4t + 6 = 3$
   $4t = -3$
   $t = -\dfrac{3}{4}$

9. $b\left(\dfrac{x}{b} + 3\right) = \dfrac{1}{b}b$
   $x + 3b = 1$
   $x = 1 - 3b$

11. $2a\left(\dfrac{y}{2a} - \dfrac{1}{a}\right) = 2(2a)$
    $y - 2 = 4a$
    $y = 4a + 2$

13. $4a^2\left(\dfrac{1}{2a} - \dfrac{3s}{4}\right) = \dfrac{1}{a^2}4a^2$
    $2a - 3a^2s = 4$
    $-3a^2s = 4 - 2a$
    $s = \dfrac{4 - 2a}{-3a^2}$

15. $4b^2\left(\dfrac{2x}{b} + \dfrac{4}{b^2}\right) = \dfrac{x}{4b}4b^2$
    $8bx + 16 = xb$
    $7bx = -16$
    $x = \dfrac{-16}{7b}$

17. $\dfrac{1}{x+2} = \dfrac{1}{2x}$
    $2x = x + 2$
    $x = 2$

19. $\dfrac{2}{3s+1} = \dfrac{1}{4}$
    $8 = 3s + 1$
    $7 = 3s$
    $s = \dfrac{7}{3}$

21. $2(x+2)\left[\dfrac{5}{2(x+2)} + \dfrac{3}{x+2}\right] = 2(2)(x+2)$
    $5 + 6 = 4x + 8$
    $11 = 4x + 8$
    $3 = 4x$
    $x = \dfrac{3}{4}$

23. $y(y-1)\left[\dfrac{1}{y(y-1)}-\dfrac{1}{y}\right]=\dfrac{1}{y-1}y(y-1)$

$1-(y-1)=y$

$1-y+1=y$

$2-y=y$

$2=2y$

$y=1$

No Solution.

25. $30\left[\dfrac{V-6}{5}+\dfrac{V-8}{15}+\dfrac{V}{10}\right]=0(30)$

$6(V-6)+2(V-8)+3V=0$

$6V-36+2V-16+3V=0$

$11V-52=0$

$11V=52$

$V=\dfrac{52}{11}$

27. $pqf\left[\dfrac{1}{p}+\dfrac{1}{q}\right]=\left[\dfrac{1}{f}\right]pqf$

$qf+pf=pq$

$qf-pq=-pf$

$q(f-p)=-pf$

$q=\dfrac{-pf}{f-p}$

29. $\dfrac{W}{Q_1}=\dfrac{T_2}{T_1}-1$

$Q_1T_1\left[\dfrac{W}{Q_1}\right]=Q_1T_1\left[\dfrac{T_2}{T_1}-1\right]$

$T_1W=Q_1T_2-Q_1T_1$

$T_1W+Q_1T_1=Q_1T_2$

$T_2=\dfrac{T_1W+Q_1T_1}{Q_1}$

31. Let x = amount of increase

$\dfrac{6000+x}{8000+x}=\dfrac{4}{5}$

$5(6000+x)=4(8000+x)$

$3000+5x=32000+4x$

$x=2000$

33. Let x = length of original section of pipe

$\dfrac{x}{3}=\dfrac{x+0.60}{4}$

$4x=3x+1.8$

$x=1.8$ mi

35. Let x = time to paint bridge working together

$\dfrac{x}{600}+\dfrac{x}{400}=1$

$1200\left[\dfrac{x}{600}+\dfrac{x}{400}\right]=1200$

$2x+3x=1200$

$x=\dfrac{1200}{5}=240$ h

Review Exercises

1. $\dfrac{9rst^4}{3s^3y^2} = \dfrac{3rt^4}{s^3}$

3. $\dfrac{2a^2bc}{6ab^2c^2} = \dfrac{a}{3bc^2}$

5. $\dfrac{4(x+2y)}{(x-2y)(x+2y)} = \dfrac{4}{(x-2y)}$

7. $\dfrac{p(p+q)}{p(3+2p^2)} = \dfrac{(p+q)}{(3+2p^2)}$

9. $\dfrac{2a(a+b-c)}{4b(a+b-c)} = \dfrac{a}{2b}$

11. $\dfrac{(2x-3y)(3x+y)}{(2x-3y)(2x-y)} = \dfrac{(3x+y)}{(2x-y)}$

13. 18

15. $12t^2$

17. $48b^2t$

19. $20x^2(x-2)$

21. $\dfrac{2x}{3a} \cdot \dfrac{5a^2}{x^2} = \dfrac{10a}{3x^2}$

23. $\dfrac{3x^2}{4y^3} \cdot \dfrac{5y^4}{x^3} = \dfrac{15y}{4x}$

25. $\dfrac{3a}{4} \cdot \dfrac{8}{a^2} = \dfrac{6}{a}$

27. $\dfrac{4u^2}{4bv^2} \cdot \dfrac{8b^2y}{a^2y} = \dfrac{2bu}{av}$

29. $\dfrac{2(5b)}{5a^2b} - \dfrac{3a}{5a^2b} = \dfrac{10b-3a}{5a^2b}$

31. $\dfrac{5(2cd)}{2c^2d} - \dfrac{3(2)}{2c^2d} + \dfrac{c}{2c^2d} = \dfrac{10cd-6+c}{2c^2d}$

33. $\dfrac{2^3}{3^3} = \dfrac{8}{27}$

35. $\dfrac{a^4x^4}{3^4(y^2)^4} = \dfrac{a^4x^4}{81y^8}$

37. $\dfrac{2x}{(x-1)(x+1)} \cdot \dfrac{x-1}{x^2} = \dfrac{2}{x(x+1)}$

39. $\dfrac{(x-5)(x+3)}{(x-3)(x+3)} \cdot \dfrac{(x-3)(x-3)}{4(x-3)} = \dfrac{x-5}{4}$

41. $\dfrac{(a-1)(a+1)}{4a} \cdot \dfrac{8a^2}{2(a+1)} = a(a-1)$

43. $\dfrac{(3x-2y)(3x+2y)}{1} \cdot \dfrac{y-2x}{3x-2y} = (3x+2y)(y-2x)$

45. $\dfrac{3}{x^2} + \dfrac{2}{x(x+3)}$

$\dfrac{3(x+3)}{x^2(x+3)} + \dfrac{2x}{x^2(x+3)}$

$\dfrac{3x+9+2x}{x^2(x+3)}$

$\dfrac{5x+9}{x^2(x+3)}$

47. $\dfrac{2x}{(x-2)} - \dfrac{x^2-3}{(x-2)(x-2)}$

$\dfrac{2x(x-2)}{(x-2)(x-2)} - \dfrac{x^2-3}{(x-2)(x-2)}$

$\dfrac{2x^2-4x-x^2+3}{(x-2)(x-2)}$

$\dfrac{x^2-4x+3}{(x-2)(x-2)} = \dfrac{(x-3)(x-1)}{(x-2)(x-2)}$

49. $\dfrac{(2x-3)}{(2x+5)(x-3)} - \dfrac{3x}{(x-3)(x+3)}$

$\dfrac{(2x-3)(x+3)}{(2x+5)(x-3)(x+3)} - \dfrac{3x(2x+5)}{(2x+5)(x-3)(x+3)}$

$\dfrac{2x^2+6x-3x-9-6x^2-15x}{(2x+5)(x-3)(x+3)}$

$\dfrac{-4x^2-12x-9}{(2x+5)(x-3)(x+3)} = \dfrac{-(4x^2+12x+9)}{(2x+5)(x-3)(x+3)}$

51. $\dfrac{3}{x(x+5)} - \dfrac{x}{(x+5)(x-5)} + \dfrac{2}{(2x-1)(x+5)}$

$\dfrac{3(x-5)(2x-1)}{x(x+5)(x-5)(2x-1)} - \dfrac{x^2(2x-1)}{x(x+5)(x-5)(2x-1)} + \dfrac{2x(x-5)}{x(x+5)(x-5)(2x-1)}$

$\dfrac{6x^2-3x-30x+15-2x^3+x^2+2x^2-10x}{x(x+5)(x-5)(2x-1)}$

$\dfrac{-2x^3+9x^2-43x+15}{x(x+5)(x-5)(2x-1)}$

53. $\dfrac{2}{(a\diagup b)} \bullet \dfrac{a\diagup b}{5} \bullet \dfrac{(a\diagup b)(a\diagup b)}{(a\diagup b)(a\diagup b)} = \dfrac{2}{5}$

55. $\dfrac{1}{(x-2)} \bullet \dfrac{(x+1)}{3+x} \bullet \dfrac{(x-3)(x\diagup 2)}{(2\diagup x)} = \dfrac{(x+1)(x-3)}{(x-2)(3+x)}$

57. $18\left[\dfrac{1}{2} - \dfrac{x}{6} + 3\right] = \dfrac{2x}{9} \bullet 18$

$9 - 3x + 54 = 4x$

$63 - 3x = 4x$

$63 = 7x$

$x = 9$

59. $4a\left[\dfrac{bx}{a} - \dfrac{1}{4}\right] = 4a\left(x + \dfrac{3}{2a}\right)$

$4bx - a = 4ax + 6$

$4bx - 4ax = 6 + a$

$x(4b - 4a) = 6 + a$

$x = \dfrac{6 + a}{4b - 4a}$

61. $x(x+1)\left[\dfrac{5}{x+1} - \dfrac{3}{x}\right] = x(x+1)\left(\dfrac{-5}{x(x+1)}\right)$

$5x - 3(x+1) = -5$

$5x - 3x - 3 = -5$

$2x = -2$

$x = -1$

No Solution

63. $4x(x+2)\left[\dfrac{5x}{x(x+2)} + \dfrac{1}{x}\right] = 4x(x+2)\left(\dfrac{3}{4(x+2)}\right)$

$20x + 4(x+2) = 3x$

$20x + 4x + 8 = 3x$

$24x + 8 = 3x$

$21x = -8$

$x = \dfrac{-8}{21}$

65. $\dfrac{3r}{12r^3} - \dfrac{h}{12r^3} = \dfrac{3r - h}{12r^3}$

67. $\dfrac{Z_1 + Z_2 - 2Z_1}{Z_1 + Z_2} = \dfrac{-Z_1 + Z_2}{Z_1 + Z_2}$

69. $\left[\dfrac{1}{U} - \dfrac{1}{h}\right] = \dfrac{x}{k}$

$\dfrac{k}{U} - \dfrac{k}{h} = x$

71. Let x = amount spent on manu

$r = 60,000 + x$

$\dfrac{5}{3} = \dfrac{60000 + x}{x}$

$5x = 180000 + 3x$

$2x = 180000$

$x = \$90,000$ manufacturing; $150,000 research

73. Let x = amount entire estate

$\dfrac{1}{5} + \dfrac{2}{3} = \dfrac{3}{15} + \dfrac{10}{15} = \dfrac{13}{15}; \dfrac{2}{15}$ remainder to last cousin

$\dfrac{2}{15}x = 25000$

$2x = 375000$

$x = 187500 \bullet \dfrac{1}{5} = \$37,500; 187500 \bullet \dfrac{2}{3} = \$125,000$

Chapter Test

1. $\dfrac{4\,m}{5\,n}$

2. $\dfrac{b\cancel{x}\left(a\diagup x\right)}{c\cancel{x}\left(a\diagup x\right)} = \dfrac{b}{c}$

3. $\dfrac{3a\left(a\diagup 2\right)}{\left(a\diagup 2\right)\left(a+2\right)} = \dfrac{3a}{\left(a+2\right)}$

4. $\dfrac{\overset{5}{\cancel{15}}\,\cancel{x}^2}{\overset{21}{7\cancel{b}}} \cdot \dfrac{\overset{4}{\cancel{28}}\,\cancel{ab}}{\underset{3}{\cancel{9}}\,\cancel{a}^3 c} = \dfrac{20}{3bc}$

5. $\dfrac{\left(5x\diagup 1\right)}{\left(x\diagup 2\right)\left(x+1\right)} \cdot \dfrac{\left(x\diagup 2\right)}{\left(5x\diagup 1\right)\left(2x+3\right)} = \dfrac{1}{\left(x+1\right)\left(2x+3\right)}$

6. $\dfrac{\overset{}{\cancel{5}}\,\cancel{xy}}{\left(a\diagup x\right)} \cdot \dfrac{\left(a\diagup x\right)\left(a+x\right)}{\underset{2}{\cancel{10}}\,\cancel{xy}} = \dfrac{\left(a+x\right)}{2}$

7. $\dfrac{3}{4x^2} - \dfrac{2}{x\left(x-1\right)} - \dfrac{x}{2\left(x-1\right)}$

   $\dfrac{3\left(x-1\right)}{4x^2\left(x-1\right)} - \dfrac{2\left(4\right)x}{4x^2\left(x-1\right)} - \dfrac{x\left(2x^2\right)}{4x^2\left(x-1\right)}$

   $\dfrac{-2x^3 - 5x - 3}{4x^2\left(x-1\right)}$

8. $\dfrac{4x}{\left(x+1\right)\left(x+2\right)} + \dfrac{7x-1}{\left(x+5\right)\left(x-1\right)}$

   $\dfrac{4x\left(x+5\right)\left(x-1\right)}{\left(x+1\right)\left(x+2\right)\left(x+5\right)\left(x-1\right)} + \dfrac{\left(7x-1\right)\left(x+1\right)\left(x+2\right)}{\left(x+1\right)\left(x+2\right)\left(x+5\right)\left(x-1\right)}$

   $\dfrac{11x^3 + 36x^2 - 9x - 2}{\left(x+1\right)\left(x+2\right)\left(x+5\right)\left(x-1\right)}$

9. $x\left(2x-3\right)\left[\dfrac{3}{x\left(2x-3\right)} + \dfrac{3}{x}\right] = \dfrac{3}{2x-3}x\left(2x-3\right)$

   $3 + 6x - 9 = 3x$

   $3x = 6$

   $x = 2$

10. Let x = time to clean dumpsite together

    $1800\left[\dfrac{x}{450} + \dfrac{x}{600}\right] = 1\left(1800\right)$

    $4x + 3x = 1800$

    $7x = 1800$

    $x = 257.1 \approx 258$ h

Chapter 10
Exponents, Roots, and Radicals

10.1 Integral Exponents

1. $\dfrac{1}{t^5}$

3. $\dfrac{1}{x^4}$

5. x^3

7. $R_1^{\ 3}$

9. $\dfrac{3}{c^2}$

11. $\dfrac{c}{3}$

13. 1

15. 1

17. 5

19. $9y^2$

21. $3 \bullet 3^5 = 3^6$

23. $\dfrac{6^5}{6^4} = \dfrac{6^{5-4}}{1} = 6$

25. $\dfrac{a^2}{xa^3} = \dfrac{1}{xa}$

27. $2c^{-8} = \dfrac{2}{c^8}$

29. $\dfrac{y^3}{xx^5y} = \dfrac{y^2}{x^6}$

31. $\dfrac{xx}{5^2 \bullet 5} = \dfrac{x^2}{5^3} = \dfrac{x^2}{125}$

33. $\dfrac{s}{t^2}$

35. $\dfrac{x^6}{2 \bullet 2^2 x^2 y^4} = \dfrac{x^4}{8y^4}$

37. $\dfrac{b^5 b^2}{3a \bullet 3} = \dfrac{b^7}{9a}$

39. $\dfrac{2^2 a^{-2}}{5^2 b^2} = \dfrac{2^2}{5^2 a^2 b^2} = \dfrac{4}{25a^2 b^2}$

41. $\dfrac{x^{-2} y^2}{x^2 y^{-3}} = \dfrac{y^2 y^3}{x^2 x^2} = \dfrac{y^5}{x^4}$

43. $\dfrac{3^{-1} a^2 b^{-1} c}{6a^{-3} b^{-1} c^{-1}} = \dfrac{a^2 a^3 cc}{6 \bullet 3} = \dfrac{a^5 c^2}{18}$

45. $R_1^{-1} + R_2^{-1} = \dfrac{1}{R_1} + \dfrac{1}{R_2}$

47. $g(cm)^{-3}; ms^{-2}$

49. $\dfrac{362 \text{ Btu}}{h \bullet ft^2}$

10.2 Fractional Exponents

1. $\sqrt{5}$

3. $\sqrt[4]{a}$

5. $\sqrt[5]{x^3}$

7. $\sqrt[3]{R^7}$

9. $a^{1/3}$

11. $x^{1/2}$

13. $x^{2/3}$

15. $b^{8/5}$

17. $\sqrt{9} = 3$

19. $\sqrt[4]{16} = \sqrt[4]{2 \bullet 2 \bullet 2 \bullet 2} = 2$

21. $\sqrt{4^3} = \sqrt{2^2 \bullet 2^2 \bullet 2^2} = 2 \bullet 2 \bullet 2 = 8$

23. $\sqrt[3]{8^4} = \sqrt[3]{8 \bullet 8 \bullet 8 \bullet 8} = \sqrt[3]{2^3 \bullet 2^3 \bullet 2^3 \bullet 2^2} = 2 \bullet 2 \bullet 2 \bullet 2 = 16$

25. $\sqrt[4]{81^3} = \sqrt[4]{3^4 \bullet 3^4 \bullet 3^4} = 3 \bullet 3 \bullet 3 = 27$

27. $\sqrt[3]{(-8)^2} = \sqrt[3]{64} = \sqrt[3]{2^3 \bullet 2^3} = 2 \bullet 2 = 4$

29. $\sqrt{36^{-1}} = \sqrt{\left(\dfrac{1}{36}\right)} = \dfrac{\sqrt{1}}{\sqrt{36}} = \dfrac{1}{6}$

31. $\sqrt[4]{\dfrac{1}{16}} = \sqrt[4]{\dfrac{1}{2^4}} = \dfrac{1}{\sqrt[4]{2^4}} = \dfrac{1}{2}$

33. $-\sqrt[5]{-\dfrac{1}{32}} = -\sqrt[5]{-\dfrac{1}{2^5}} = -\dfrac{-1}{2} = \dfrac{1}{2}$

35. $\sqrt{4} \bullet \sqrt[3]{27^2} = 2\sqrt[3]{3^3 \bullet 3^3} = 2 \bullet 3 \bullet 3 = 18$

37. $2^{-1/2 + 3/2} = 2^{2/2} = 2$

39. $a^{2/3 + 1/2} = a^{3/2}$

41. $a^{4/4 - 1/4}b = a^{3/4}b$

43. $\dfrac{x^{3/5}x^{4/5}}{x^{2/5}} = \dfrac{x^{7/5}}{x^{2/5}} = x^{7/5 - 2/5} = x^{35/15 - 6/15} = x^{29/15}$

45. $\dfrac{x^{-1/2 + 2}}{x^{2/3}} = \dfrac{x^{3/2}}{x^{2/3}} = x^{3/2 - 2/3} = x^{9/6 - 4/6} = x^{5/6}$

47. $\begin{aligned} \sqrt{x^5} + \sqrt{y^5} &= k \\ x^{5/2} + y^{5/2} &= k \end{aligned}$

49. $R = \dfrac{\left(1 + D_1^2\right)^{3/2}}{D_2}$

$R = \dfrac{\sqrt{\left(1 + D_1^2\right)^3}}{D_2}$

51. $9.60 \bullet 10^{18} T^{3/2}$

$9.60 \bullet 10^{18} \bullet 289 = 4.72 \bullet 10^{22}$

10.3 Imaginary Roots

1. 7

3. −12

5. 0.4

7. −0.2

9. 20

11. −40

13. $\sqrt[3]{8} = 2$

15. $\sqrt[3]{-8} = -2$

17. $-\sqrt[3]{125} = -5$

19. $\sqrt[3]{0.125} = 0.5$

21. $\sqrt[4]{16} = 2$

23. $\sqrt[5]{243} = 3$

25. $\sqrt[6]{64} = 2$

27. $\sqrt{-4} = 2\sqrt{-1} = 2j$

29. $-\sqrt{-400} = -20\sqrt{-1} = -20j$

31. $\sqrt{-0.49} = 0.7\sqrt{-1} = 0.7j$

33. $(5j)^2 = 5^2 j^2 = 25\sqrt{-1}\sqrt{-1} = 25(-1) = -25$

35. $(5j)(4j) = 20j^2 = 5\sqrt{-1}\sqrt{-1} = 20(-1) = -20$

37. $(-3j)(-4j) = 12j^2 = 12\sqrt{-1}\sqrt{-1} = 12(-1) = -12$

39. $-(8)^2 j^2 = -64\sqrt{-1}\sqrt{-1} = -64(-1) = -(-64) = 64$

41. $e^3 = 216$

$\sqrt[3]{e} = \sqrt[3]{216}$

$e = 6.00 \text{ ft}$

43. $\dfrac{r_1}{r_2} = \dfrac{\sqrt{m_2}}{\sqrt{m_1}}$

$\dfrac{r_1}{r_2} = \sqrt{\dfrac{81}{25}} = \dfrac{9}{5}$

45. $\sqrt[3]{\dfrac{V}{I}} - 1$

$\sqrt[3]{\dfrac{532.40}{400}} - 1 = 0.1$

10.4 Simplifying Radicals

1. $\dfrac{1}{\sqrt{2}} \bullet \dfrac{\sqrt{2}}{\sqrt{2}} = \dfrac{\sqrt{2}}{2}$

3. $\dfrac{2}{\sqrt{5}} \bullet \dfrac{\sqrt{5}}{\sqrt{5}} = \dfrac{2\sqrt{5}}{5}$

5. $\dfrac{1}{\sqrt{a}} \bullet \dfrac{\sqrt{a}}{\sqrt{a}} = \dfrac{\sqrt{a}}{a}$

7. $\dfrac{\sqrt{a}}{\sqrt{b}} \bullet \dfrac{\sqrt{b}}{\sqrt{b}} = \dfrac{\sqrt{ab}}{b}$

9. $\sqrt{\dfrac{3}{5}} = \dfrac{\sqrt{3}}{\sqrt{5}} \cdot \dfrac{\sqrt{5}}{\sqrt{5}} = \dfrac{\sqrt{15}}{5}$

11. $\sqrt{\dfrac{a^3}{3}} = \dfrac{\sqrt{a^3}}{\sqrt{3}} \cdot \dfrac{\sqrt{3}}{\sqrt{3}} = \dfrac{\sqrt{3a^3}}{3} = \dfrac{a\sqrt{3a}}{3}$

13. $\sqrt{12} = \sqrt{4 \cdot 3} = 2\sqrt{3}$

15. $\sqrt{28} = \sqrt{4 \cdot 7} = 2\sqrt{7}$

17. $\sqrt{45} = \sqrt{9 \cdot 5} = 3\sqrt{5}$

19. $\sqrt{150} = \sqrt{25 \cdot 6} = 5\sqrt{6}$

21. $\sqrt{147} = \sqrt{49 \cdot 3} = 7\sqrt{3}$

23. $\sqrt{243} = \sqrt{81 \cdot 3} = 9\sqrt{3}$

25. $\sqrt{ac^2} = c\sqrt{a}$

27. $\sqrt{a^3b^2} = ab\sqrt{a}$

29. $\sqrt{4a^2bc^3} = 2ac\sqrt{bc}$

31. $\sqrt{80x^4yz^5} = \sqrt{16 \cdot 5x^4yz^5} = 4x^2z^2\sqrt{5yz}$

33. $\dfrac{\sqrt{ab^2}}{\sqrt{12}} \cdot \dfrac{\sqrt{12}}{\sqrt{12}} = \dfrac{\sqrt{12ab^2}}{12} = \dfrac{\sqrt{4 \cdot 3ab^2}}{12} = \dfrac{\overset{}{2}b\sqrt{3a}}{\underset{6}{12}} = \dfrac{b\sqrt{3a}}{6}$

35. $\dfrac{\sqrt{2x^2y}}{\sqrt{5a^8}} \cdot \dfrac{\sqrt{5a^8}}{\sqrt{5a^8}} = \dfrac{\sqrt{10x^2ya^8}}{5a^8} = \dfrac{xa^4\sqrt{10y}}{5a^{8\,4}} = \dfrac{x\sqrt{10y}}{5a^4}$

37. $\sqrt[3]{54} = \sqrt[3]{3 \cdot 3 \cdot 3 \cdot 2} = 3\sqrt[3]{2}$

39. $\sqrt[3]{8a^4} = \sqrt[3]{2 \cdot 2 \cdot 2a^4} = 2a\sqrt[3]{a}$

41. $\sqrt[4]{16a^9} = \sqrt[4]{2 \cdot 2 \cdot 2 \cdot 2a^9} = 2a^2\sqrt[4]{a}$

43. $\sqrt[4]{243a^{11}} = \sqrt[4]{3 \cdot 3 \cdot 3 \cdot 3 \cdot 3a^{11}} = 3a^2\sqrt[4]{3a^3}$

45. $\sqrt[4]{162a^{10}x^{12}} = \sqrt[4]{3 \cdot 3 \cdot 3 \cdot 3 \cdot 2a^{10}x^{12}} = 3a^2x^3\sqrt[4]{2a^2}$

47. $\sqrt[7]{256r^7s^{14}t^{16}} = \sqrt[7]{2^7 \cdot 2r^7s^{14}t^{16}} = 2rs^2t^2\sqrt[7]{2t^2}$

49. $d = 2\sqrt{\dfrac{A}{3.14}}$

$d = \dfrac{2\sqrt{A}}{\sqrt{3.14}} \cdot \dfrac{\sqrt{3.14}}{\sqrt{3.14}} = \dfrac{2\sqrt{3.14A}}{3.14}$

51. Let the width of the rectangle be w, then the length is 2w.

$12800 = 2w(w)$

$12800 = 2w^2$

$6400 = w^2$

$\sqrt{6400} = \sqrt{w^2}$

80 ft = width

160 ft = length

$$53. \quad V = k\sqrt[3]{\dfrac{P}{W}}$$

$$V = \dfrac{k\sqrt[3]{P}}{\sqrt[3]{W}} \cdot \dfrac{\sqrt[3]{W^2}}{\sqrt[3]{W^2}} = \dfrac{k\sqrt[3]{PW^2}}{W}$$

$$55. \quad d = k\sqrt[3]{\dfrac{16J}{C}}$$

$$d = \dfrac{k\sqrt[3]{16J}}{\sqrt[3]{C}} \cdot \dfrac{\sqrt[3]{C^2}}{\sqrt[3]{C^2}} = \dfrac{2k\sqrt[3]{2JC^2}}{C}$$

10.5 Operations with Radicals

1. $2\sqrt{7}$

3. $4\sqrt{7} + \sqrt{5}$

5. $2\sqrt{3} + 3\sqrt{3} = 5\sqrt{3}$

7. $2 \bullet 2\sqrt{10} + 5\sqrt{10} = 9\sqrt{10}$

9. $\sqrt{2 \bullet 50} = \sqrt{2 \bullet 2 \bullet 5 \bullet 5} = 2 \bullet 5 = 10$

11. $\sqrt{5 \bullet 15} = \sqrt{5 \bullet 5 \bullet 3} = 5\sqrt{3}$

13. $2\sqrt{2} + 3\sqrt{2} + 4\sqrt{2} = 9\sqrt{2}$

15. $2\sqrt{7} - 2 \bullet 3\sqrt{7} + 5\sqrt{7} = (2 - 6 + 5)\sqrt{7} = \sqrt{7}$

17. $2 \bullet 2\sqrt{2} - 2 \bullet 2\sqrt{3} - 5 \bullet \sqrt{2} = (4 - 5)\sqrt{2} - 4\sqrt{3} = -\sqrt{2} - 4\sqrt{3}$

19. $\sqrt{a} + 3\sqrt{a} = (3 + 1)\sqrt{a} = 4\sqrt{a}$

21. $\sqrt{2a} + 2\sqrt{2a} + 4a\sqrt{2a} = (1 + 2 + 4a)\sqrt{2a} = (3 + 4a)\sqrt{2a}$

23. $a\sqrt{2} + 6a\sqrt{2} - 2\sqrt{3a} = (a + 6a)\sqrt{2} - 2\sqrt{3a} = 7a\sqrt{2} - 2\sqrt{3a}$

25. $\sqrt{21} - 3\sqrt{9 \bullet 2}$

$\sqrt{21} - 3 \bullet 3\sqrt{2}$

$\sqrt{21} - 9\sqrt{2}$

27. $\sqrt{2 \bullet 8} - \sqrt{32 \bullet 2} + 5\sqrt{18 \bullet 2}$

$\sqrt{2 \bullet 2 \bullet 2 \bullet 2} - \sqrt{16 \bullet 2 \bullet 2} + 5\sqrt{9 \bullet 2 \bullet 2}$

$2 \bullet 2 - 4 \bullet 2 + 5 \bullet 3 \bullet 2 = 26$

29. $\sqrt{a^2 b} + 3\sqrt{a^2 c}$

$a\sqrt{b} + 3a\sqrt{c}$

31. $2\sqrt{4} - \sqrt{6} + 2\sqrt{6} - \sqrt{9}$

$2 \bullet 2 - \sqrt{6} + 2\sqrt{6} - 3$

$1 + \sqrt{6}$

33. $2\sqrt{25} + \sqrt{27 \bullet 5} - 6\sqrt{15} - 3\sqrt{81}$

$2 \bullet 5 + \sqrt{9 \bullet 3 \bullet 5} - 6\sqrt{15} - 3 \bullet 9$

$10 + 3\sqrt{15} - 6\sqrt{15} - 27$

$-17 - 3\sqrt{15}$

35. $2a + 5\sqrt{ac} - 6\sqrt{ac} - 15c$

$2a - \sqrt{ac} - 15c$

37. $\left(\sqrt{3}\sqrt{3}\right) + 2\sqrt{3} + 2\sqrt{3} + 2(2)$

$3 + 4\sqrt{3} + 4$

$7 + 4\sqrt{3}$

39. $\dfrac{\sqrt{3}+\sqrt{2}}{\sqrt{2}} \bullet \dfrac{\sqrt{2}}{\sqrt{2}} = \dfrac{\sqrt{3 \bullet 2}+\sqrt{2 \bullet 2}}{\sqrt{2 \bullet 2}} = \dfrac{\sqrt{6}+2}{2}$

41. $\dfrac{2\sqrt{2}+5\sqrt{2}}{\sqrt{2}} \bullet \dfrac{\sqrt{2}}{\sqrt{2}} = \dfrac{2\sqrt{2 \bullet 2}+5\sqrt{2 \bullet 2}}{\sqrt{2 \bullet 2}} = \dfrac{2 \bullet 2 + 5 \bullet 2}{2} = \dfrac{14}{2} = 7$

43. $\dfrac{\sqrt{7}}{\sqrt{3}+\sqrt{7}} \bullet \dfrac{\sqrt{3}-\sqrt{7}}{\sqrt{3}-\sqrt{7}} = \dfrac{\sqrt{7 \bullet 3}-\sqrt{7 \bullet 7}}{3-7} = \dfrac{\sqrt{21}-7}{-4}$

45. $\dfrac{\sqrt{3}+\sqrt{6}}{2\sqrt{3}-\sqrt{6}} \bullet \dfrac{2\sqrt{3}+\sqrt{6}}{2\sqrt{3}+\sqrt{6}} = \dfrac{6+\sqrt{18}+2\sqrt{18}+6}{2 \bullet 2 \bullet 3 - 6} = \dfrac{12+3\sqrt{2}+6\sqrt{2}}{6} = \dfrac{12+9\sqrt{2}}{6} = \dfrac{3\left(4+3\sqrt{2}\right)}{6} = \dfrac{4+3\sqrt{2}}{2}$

47. $\dfrac{\sqrt{a}}{\sqrt{a}+2\sqrt{b}} \bullet \dfrac{\sqrt{a}-2\sqrt{b}}{\sqrt{a}-2\sqrt{b}} = \dfrac{a-2\sqrt{ab}}{a-4b}$

49. $\dfrac{\sqrt{R_1 R_2}}{\sqrt{R_1}+\sqrt{R_2}} \bullet \dfrac{\sqrt{R_1}-\sqrt{R_2}}{\sqrt{R_1}-\sqrt{R_2}} = \dfrac{R_1\sqrt{R_2}-R_2\sqrt{R_1}}{R_1-R_2}$

51. $\dfrac{\sqrt{10}-\sqrt{3}}{\sqrt{10}+\sqrt{3}} \bullet \dfrac{\sqrt{10}-\sqrt{3}}{\sqrt{10}-\sqrt{3}} = \dfrac{10-\sqrt{30}-\sqrt{30}+3}{10-3} = \dfrac{13-2\sqrt{30}}{7}$

10.6 Working with Radicals

1. $\left(\sqrt{x-8}\right)^2 = 2^2$

$x-8=4$

$x=12$

3. $\left(8-2x\right)^2 = 4^2$

$8-2x=16$

$-2x=8$

$x=-4$

5. $\sqrt{3-x} = \dfrac{6}{2}$

$\left(\sqrt{3-x}\right)^2 = 3^2$

$3-x=9$

$-x=6$

$x=-6$

7. $-3\sqrt{2x+1} = -9$

$\sqrt{2x+1} = 3$

$\left(\sqrt{2x+1}\right)^2 = 3^2$

$2x+1=9$

$2x=8$

$x=4$

9. $\left(\sqrt[4]{5-x}\right)^4 = 2^4$

$5 - x = 16$

$-x = 11$

$x = -11$

11. $6 = \sqrt{(3+x)(3+x)}$

$6^2 = \left(\sqrt{(3+x)(3+x)}\right)^2$

$36 = (3+x)(3+x)$

$36 = 9 + 6x + x^2$

$0 = x^2 + 6x - 27$

$0 = (x-3)(x+9)$

$x = 3$

13. $\left(\sqrt{2x+1}\right)^2 = \left(\dfrac{9}{3\sqrt{2x+1}}\right)^2$

$2x+1 = \dfrac{9}{2x+1}$

$9 = (2x+1)(2x+1)$

$9 = 4x^2 + 4x + 1$

$0 = 4x^2 + 4x - 8$

$0 = 4(x-1)(x+4)$

$x = 1$

15. $\left(\sqrt{x+1}\right)^2 = \left[1 + \sqrt{x+4}\right]^2$

$x+1 = 1 + \sqrt{x+4} + \sqrt{x+4} + x + 4$

$x + 1 - 1 - 4 - x = 2\sqrt{x+4}$

$-4 = 2\sqrt{x+4}$

$(-2) = \left(\sqrt{x+4}\right)$

No solution

17. $f = \dfrac{1}{2\pi\sqrt{LC}}$

$2\pi f \sqrt{LC} = 1$

$\left(\sqrt{LC}\right)^2 = \left(\dfrac{1}{2\pi f}\right)^2$

$LC = \dfrac{1}{4\pi^2 f^2}$

$L = \dfrac{1}{4\pi^2 f^2 C}$

19. $r^2 = \left(\sqrt{\dfrac{A}{\pi}}\right)^2$

$r^2 = \dfrac{A}{\pi}$

$\pi r^2 = A$

21. $v = \sqrt{v_0^2 - 2gh}$

$v^2 = v_0^2 - 2gh$

$v^2 - v_0^2 = -2gh$

$\dfrac{v^2 - v_0^2}{-2g} = h$

23. $v = \sqrt{usgR}$

$v^2 = usgR$

$\dfrac{v^2}{usg} = R$

25. $S = \pi r \sqrt{r^2 + h^2}$

$\dfrac{S}{\pi r} = \sqrt{r^2 + h^2}$

$\dfrac{S^2}{\pi^2 r^2} = r^2 + h^2$

$\dfrac{S^2}{\pi^2 r^2} - r^2 = h^2$

$\sqrt{\dfrac{S^2}{\pi^2 r^2} - r^2} = h$

Chapter Review

1. $\dfrac{1}{10}$

3. $3^2 = 9$

5. $\dfrac{1}{6}$

7. 13

9. 5

11. $\dfrac{1}{4}$

13. $\dfrac{-3}{11}$

15. $\sqrt{100} = 10$

17. $\sqrt{49^3} = \sqrt{49 \bullet 49 \bullet 49} = 7 \bullet 7 \bullet 7 = 343$

19. $\sqrt[3]{8^7} = 128$

21. $\sqrt{25^3} \bullet 5 = \sqrt{5^2 \bullet 5^2 \bullet 5^2} \bullet 5 = 5 \bullet 5 \bullet 5 \bullet 5 = 625$

23. $\sqrt{-144} = 12\sqrt{-1} = 12j$

25. $-\sqrt{-0.01} = -0.1\sqrt{-1} = -0.1j$

27. $\dfrac{2x}{y}$

29. $\dfrac{2rt^5}{s}$

31. $\dfrac{cc^3}{4ba^2 b} = \dfrac{c^4}{4b^2 a^2}$

33. $x^{\frac{1}{5} + \frac{2}{3}} = x^{\frac{3}{15} + \frac{10}{15}} = x^{\frac{13}{15}}$

35. $\dfrac{b^{\frac{1}{2}}}{b} = b^{\frac{1}{2} - \frac{2}{2}} = b^{-\frac{1}{2}} = \dfrac{1}{b^{\frac{1}{2}}}$

37. $8^{\frac{1}{3}} x^{-1} y^{(\frac{1}{2})(\frac{2}{3})} = \dfrac{2y^{\frac{1}{2}}}{x}$

39. $\dfrac{16^{\frac{3}{4}} c^{\frac{9}{4}} c^{(\frac{1}{5})}}{a} = \dfrac{8c^{\frac{30}{20} + \frac{4}{22}}}{a} = \dfrac{8c^{\frac{34}{20}}}{a} = \dfrac{8c^{\frac{17}{10}}}{a}$

41. $3\sqrt{3}$

43. $\sqrt{9 \bullet 6} = 3\sqrt{6}$

45. $\sqrt{4 \bullet 31} = 2\sqrt{31}$

47. $\sqrt[3]{3 \bullet 3 \bullet 3 \bullet 4} = 3\sqrt[3]{4}$

49. $\dfrac{\sqrt{4}}{\sqrt{7}} \cdot \dfrac{\sqrt{7}}{\sqrt{7}} = \dfrac{\sqrt{4 \cdot 7}}{7} = \dfrac{2\sqrt{7}}{7}$

51. $\sqrt{4 \cdot 7a} = 2\sqrt{7a}$

53. $\sqrt{9 \cdot 10b^3} = 3b\sqrt{10b}$

55. $\dfrac{\sqrt{3}}{a}$

57. $\dfrac{\sqrt{400000}}{\sqrt{ab}} \cdot \dfrac{\sqrt{ab}}{\sqrt{ab}} = \dfrac{\sqrt{40000 \cdot 10ab}}{ab} = \dfrac{200\sqrt{10ab}}{ab}$

59. $\sqrt[3]{3 \cdot 3 \cdot 3 \cdot 3x^2 y^4} = 3y\sqrt[3]{3x^2 y}$

61. $6\sqrt{5} - 12\sqrt{5} = -6\sqrt{5}$

63. $a\sqrt{2} + 6\sqrt{2} - 4a\sqrt{2} = (6 - 3a)\sqrt{2}$

65. $\sqrt{12} - 2\sqrt{24 \cdot 2}$

   $\sqrt{4 \cdot 3} - 2\sqrt{2 \cdot 2 \cdot 2 \cdot 3 \cdot 2}$

   $2\sqrt{3} - 2 \cdot 2 \cdot 2\sqrt{3}$

   $-6\sqrt{3}$

67. $a\sqrt{b} - 3\sqrt{5ab}$

69. $12 - 3\sqrt{30} - 2\sqrt{30} + 15$

   $27 - 5\sqrt{30}$

71. $2ab + \sqrt{abc} - 2\sqrt{abc} - c$

   $2ab - \sqrt{abc} - c$

73. $\dfrac{\sqrt{2} - 1}{2\sqrt{2} + 3} \cdot \dfrac{\left(2\sqrt{2} - 3\right)}{\left(2\sqrt{2} - 3\right)} = \dfrac{2\sqrt{2 \cdot 2} - 3\sqrt{2} - 2\sqrt{2} + 3}{8 - 9} = \dfrac{4 - 3\sqrt{2} - 2\sqrt{2} + 3}{-1} = \dfrac{7 - 5\sqrt{2}}{-1} = -7 + 5\sqrt{2}$

75. $\dfrac{r^2 + 4R^2}{2\pi R}$

77. $v = \dfrac{\sqrt{2eV}}{\sqrt{m}} \cdot \dfrac{\sqrt{m}}{\sqrt{m}} = \dfrac{\sqrt{2eVm}}{m}$

79. $\dfrac{a}{\sqrt{2}} \cdot \dfrac{2}{a\sqrt{6}} = \dfrac{2a}{a\sqrt{12}} \cdot \dfrac{\sqrt{12}}{\sqrt{12}} = \dfrac{2a\sqrt{12}}{12a} = \dfrac{4a\sqrt{3}}{12a} = \dfrac{\sqrt{3}}{3}$

81. $1000\sqrt[4]{w}$

   $1000\sqrt[4]{25} = \text{thermo dym temp}$

   $2236.07 - 273 = 1963.07 \quad C$

83. $\left(\sqrt{3x^2 - 8}\right)^2 = x^2$

   $3x^2 - 8 = x^2$

   $2x^2 = 8$

   $x^2 = 4$

   $x = 2$

85. $2x - 1 = x + 3$

   $x = 4$

Chapter Test

1. $4\sqrt{5} - 5\sqrt{5} = -\sqrt{5}$

2. $\sqrt{100 \bullet 100 \bullet 100} \bullet \sqrt[3]{8 \bullet 8} = 10 \bullet 10 \bullet 10 \bullet 2 \bullet 2 = 4{,}000$

3. $\sqrt{9 \bullet 3a^4b^3} = 3a^2b\sqrt{3b}$

4. $\dfrac{2b^{3/4}b^{3/4}}{a^{1/2}ba^{2/2}} = \dfrac{2b^{11/4 - 1/4}}{a^{1/2 + 2/2}} = \dfrac{2b^{7/4}}{a^{3/2}}$

5. $\left(\sqrt{2x} - 3\sqrt{y}\right)\left(\sqrt{2x} - 3\sqrt{y}\right)$

   $2x - 3\sqrt{2xy} - 3\sqrt{2xy} + 9y$

   $2x - 6\sqrt{2xy} + 9y$

6. $\dfrac{3 - 2\sqrt{x}}{2\sqrt{x}} \bullet \dfrac{\sqrt{x}}{\sqrt{x}} = \dfrac{3\sqrt{x} - 2x}{2x}$

7. $\dfrac{2\sqrt{15x} + \sqrt{3}}{\sqrt{15x} - 2\sqrt{3}} \bullet \dfrac{\sqrt{15x} + 2\sqrt{3}}{\sqrt{15x} + 2\sqrt{3}} = \dfrac{30x + 4\sqrt{15 \bullet 3x} + \sqrt{3 \bullet 15x} + 6}{15x - 4 \bullet 3}$

   $\dfrac{30x + 15\sqrt{5x} + 6}{15x - 12} = \dfrac{3\left(10x + 5\sqrt{5x} + 2\right)}{3\left(5x - 4\right)} = \dfrac{\left(10x + 5\sqrt{5x} + 2\right)}{5x - 4}$

8. $\left(\sqrt{x^2 - 3x}\right)^2 = (x - 7)^2$

   $x^2 - 3x = x^2 - 14x + 49$

   $11x = 49$

   $x = \dfrac{49}{11}$

9. $v = k\left(\sqrt[3]{\dfrac{P}{W}}\right)$

   a) $k\left(\dfrac{P}{W}\right)^{1/3}$

   b) $k\dfrac{\sqrt[3]{PW^2}}{W}$

10. $\left(\dfrac{2n}{\omega r}\right)^{-1/2}$

   $\dfrac{(\omega r)^{1/2}}{(2n)^{1/2}}$

   $\dfrac{\sqrt{\omega r}}{\sqrt{2n}} \bullet \dfrac{\sqrt{2n}}{\sqrt{2n}} = \dfrac{\sqrt{2n\omega r}}{2n}$

Chapter 11
Quadratic Equations

11.1 The Quadratic Equation

1. yes; $x^2 - 7x - 4 = 0; a = 1, b = -7, c = -4$

3. no; $-2x + 1 = 0$;

5. yes; $x^2 + 2x + 4 = 0; a = 1, b = 2, c = 4$

7. yes; $7x^2 - x = 0; a = 7, b = -1, c = 0$

9. no; $x^2 - x^3 = 0$

11. no; $3x^3 - 9x^2 + 3x - 1 = 0$

13. $x = 2 : 4 - 10 + 6 = 0$
    $x = 3 : 9 - 15 + 6 = 0$

15. $x = 2 : 4 - 8 + 4 = 0$

17. $x = -1 : 1 + 1 - 2 = 0$
    $x = 2 : 4 - 2 - 2 = 0$

19. $x = \dfrac{1}{2} : \dfrac{1}{2} - \dfrac{3}{2} + 1 = 0$
    $x = 1 : 2 - 3 + 1 = 0$

21. $v = 3 : 9 + 3 = 12$
    $v = -4 : 16 - 4 = 12$

23. none

25. $5x^2 - 8x - 132 = 0$;
    $5(36) - 48 - 132 = 0$
    $180 - 48 - 132 = 0$

27. $500r^2 + 1000r - 105 = 0$;
    $500(0.01) + 1000(0.1) - 105 = 0$
    $5 + 100 - 105 = 0$

11.2 Solving Quadratic Equations by Factoring

1. $x = 3, x = -3$;
   $(x - 3)(x + 3) = 0$

3. $x = -2, x = 1$;
   $(x + 2)(x - 1) = 0$

5. $t = 5, t = -2$;
   $(t - 5)(t + 2) = 0$

7. $x = \dfrac{1}{3}, x = -2$;
   $(3x - 1)(x + 2) = 0$

9. $x = \dfrac{1}{2}, x = -2$;
   $(2x - 1)(x + 2) = 0$

11. $x = \dfrac{-5}{2}, x = \dfrac{1}{3}$;
    $(2x + 5)(3x - 1) = 0$

13. $n = \dfrac{2}{3}, n = \dfrac{-1}{2}$;
    $(3n - 2)(2n + 1) = 0$

15. $x = \dfrac{1}{5}, x = 4$;
    $(5x - 1)(x - 4) = 0$

17. $x = \dfrac{9}{2}, x = -1$;
    $(2x - 9)(x + 1) = 0$

19. $x = -2$ double root;
$(x+2)(x+2) = 0$

21. $R = 0, R = 8$;
$R(R-8) = 0$

23. $m = \frac{1}{2}, m = \frac{-7}{2}$;
$2(2m-1)(2m+7) = 0$

25. $x = \frac{7}{2}$ double root;
$(2x-7)(2x-7) = 0$

27. $R = 0, R = 7$;
$R(R-7) = 0$

29. $m = \frac{-1}{3}, m = 3$;
$(3m+1)(m-3) = 0$

31. $x = 2a, x = -2a$;
$(x-2a)(x+2a) = 0$

33. $x = 17, x = 2$;
$-1(x-17)(x-2) = 0$

35. $x = 25, x = 10$;
$x(35-x) = 250$
$x^2 - 35x + 250 = 0$
$(x-25)(x-10) = 0$

37. $t = 2.5$ sec;
$100 = 16t^2$
$16t^2 - 100 = 0$
$4(2t-5)(2t+5) = 0$

39. $I = 16, I = 1$;
$100(I-16)(I-1) = 0$

41. $t = 9$;
$(t-9)(t+15) = 0$

11.3 Completing the Square

1. $x = -2 \pm \sqrt{13}$;
$x^2 + 4x = 9$
$x^2 + 4x + 4 = 9 + 4$
$(x+2)^2 = 13$
$x + 2 = \pm\sqrt{13}$
$x = -2 \pm \sqrt{13}$

3. $x = \dfrac{3 \pm j\sqrt{15}}{2}$;
$x^2 - 3x = -6$
$x^2 - 3x + \dfrac{9}{4} = -6 + \dfrac{9}{4}$
$\left(x - \dfrac{3}{2}\right)^2 = \dfrac{-15}{4}$
$x - \dfrac{3}{2} = \dfrac{\pm\sqrt{-15}}{2}$

5. $x = -1, x = -2$;
$x^2 + 3x = -2$
$x^2 + 3x + \dfrac{9}{4} = -2 + \dfrac{9}{4}$
$\left(x + \dfrac{3}{2}\right)^2 = \dfrac{1}{4}$
$x + \dfrac{3}{2} = \pm\dfrac{1}{2}$

7. $s = -3 \pm \sqrt{13}$;
$s^2 + 6s = 4$
$s^2 + 6s + 9 = 4 + 9$
$(s+3)^2 = 13$
$s + 3 = \pm\sqrt{13}$
$s = -3 \pm \sqrt{13}$

9. $v = 3, v = -5$;
$v^2 + 2v = 15$
$v^2 + 2v + 1 = 15 + 1$
$(v+1)^2 = 16$
$v + 1 = \pm 4$
$v = -1 \pm 4$

11. $x = -1 \pm \sqrt{\dfrac{7}{3}}$;
$3x^2 + 6x = 4, \ x^2 + 2x = \dfrac{4}{3}$
$x^2 + 2x + 1 = \dfrac{4}{3} + 1$
$(x+1)^2 = \dfrac{7}{3}$
$x + 1 = \pm\sqrt{\dfrac{7}{3}}$

13. $x = \dfrac{1 \pm \sqrt{7}}{4}$;

$8x^2 - 4x = 3,\ x^2 - \dfrac{1}{2}x = \dfrac{3}{8}$

$x^2 - \dfrac{1}{2}x + \dfrac{1}{16} = \dfrac{3}{8} + \dfrac{1}{16}$

$\left(x - \dfrac{1}{4}\right)^2 = \dfrac{7}{16}$

$x - \dfrac{1}{4} = \dfrac{\pm\sqrt{7}}{4}$

15. $x = \dfrac{1}{2},\ x = -3$;

$x^2 + \dfrac{5}{2}x = \dfrac{3}{2}$

$x^2 + \dfrac{5}{2}x + \dfrac{25}{16} = \dfrac{3}{2} + \dfrac{25}{16}$

$\left(x + \dfrac{5}{4}\right)^2 = \dfrac{49}{16}$

$x + \dfrac{5}{4} = \pm\dfrac{7}{4}$

17. $y = \dfrac{1}{2} \pm \sqrt{\dfrac{11}{12}}$;

$3y^2 - 3y = 2,\ y^2 - y = \dfrac{2}{3}$

$y^2 - y + \dfrac{1}{4} = \dfrac{2}{3} + \dfrac{1}{4}$

$\left(y - \dfrac{1}{2}\right)^2 = \dfrac{11}{12},\ y - \dfrac{1}{2} = \pm\sqrt{\dfrac{11}{12}}$

19. $x = \dfrac{-1}{3}$;

$9x^2 + 6x = -1,\ x^2 + \dfrac{2}{3}x = \dfrac{-1}{9}$

$x^2 + \dfrac{2}{3}x + \dfrac{1}{9} = \dfrac{-1}{9} + \dfrac{1}{9}$

$\left(x + \dfrac{1}{3}\right)^2 = 0$

$x + \dfrac{1}{3} = 0$

21. $T = 5,\ T = 15$;

$-0.2T^2 + 4T = 15$

$T^2 - 20T = -75$

$T^2 - 20T + 100 = -75 + 100$

$(T - 10)^2 = 25$

$T - 10 = \pm 5$

$T = 10 \pm 5$

11.4 Solving Quadratic Equations Using the Quadratic Formula

1. $x = 3,\ x = -1$;

$x = \dfrac{2 \pm \sqrt{(-2)^2 - 4(1)(-3)}}{2(1)}$

3. $x = \dfrac{-1}{2},\ x = -3$;

$x = \dfrac{-7 \pm \sqrt{7^2 - 4(2)(3)}}{2(2)}$

5. $x = \dfrac{-5 \pm \sqrt{13}}{2}$;

$x = \dfrac{-5 \pm \sqrt{5^2 - 4(1)(3)}}{2(1)}$

7. $s = 2 \pm \sqrt{6}$;

$s = \dfrac{4 \pm \sqrt{(-4)^2 - 4(1)(-2)}}{2(1)}$

9. $t = \dfrac{3}{2},\ t = \dfrac{1}{2}$;

$t = \dfrac{8 \pm \sqrt{(-8)^2 - 4(4)(3)}}{2(4)}$

11. $x = \dfrac{4}{3},\ x = \dfrac{-4}{3}$;

$x = \dfrac{0 \pm \sqrt{0^2 - 4(9)(-16)}}{2(9)}$

13. $x = \dfrac{7}{2},\ x = \dfrac{-1}{2}$;

$x = \dfrac{12 \pm \sqrt{(-12)^2 - 4(4)(-7)}}{2(4)}$

15. $R = -1 \pm 2j$;

$R = \dfrac{-2 \pm \sqrt{2^2 - 4(1)(5)}}{2(1)}$

17. $t = \dfrac{1 \pm \sqrt{33}}{2}$;

$t = \dfrac{1 \pm \sqrt{(-1)^2 - 4(1)(-8)}}{2(1)}$

19. $x = \dfrac{-1 \pm \sqrt{33}}{4}$;

$x = \dfrac{-2 \pm \sqrt{2^2 - 4(4)(-8)}}{2(4)}$

21. $I = 7, I = 0$;

$I = \dfrac{7 \pm \sqrt{(-7)^2 - 4(1)(0)}}{2(1)}$

23. $t = \dfrac{3 \pm \sqrt{73}}{4}$;

$t = \dfrac{3 \pm \sqrt{(-3)^2 - 4(2)(8)}}{2(2)}$

25. $t = -\dfrac{2}{3}, t = 4$;

$t = \dfrac{10 \pm \sqrt{(-10)^2 - 4(3)(-8)}}{2(3)}$

27. $x = -\dfrac{1}{3a}, x = -\dfrac{3}{2a}$;

$x = \dfrac{-11a \pm \sqrt{(11a)^2 - 4(6a^2)(3)}}{2(6a^2)}$

29. 16 & 17;

$x(x+1) = 272$

31. 17 cm by 12 cm;

$x(29 - x) = 204$

33. $P = 0.38$;

$P = \dfrac{3 \pm \sqrt{(-3)^2 - 4(1)(1)}}{2(1)}$

35. 64.9 m;

$(400 - 2x)(500 - 2x) = \dfrac{1}{2}(400)(500)$

37. $x = 17$;

$50 = 0.1x^2 + 0.8x + 7$

$0 = 0.1x^2 + 0.8x - 43$

$0 = 1x^2 + 8x - 430$

11.5 Graphing the Quadratic Function

1. up, minimum; $a > 0$

3. down, maximum; $a < 0$

5. up, minimum; $a > 0$

7. vertex = $(3, -4)$,
   y-intercept = $(0, 5)$

9. vertex = $(\dfrac{5}{3}, \dfrac{13}{3})$,
   y-intercept = $(0, -4)$

11. vertex = $(-2, -13)$,
    y-intercept = $(0, -5)$

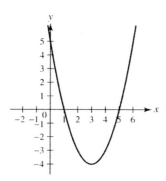

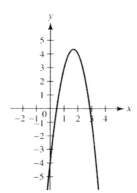

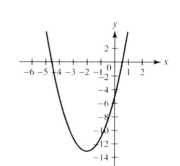

13. vertex = $(0, -4)$,
     y-intercept = $(0, -4)$,
     x-intercepts = $(2, 0)$ $(-2, 0)$

15. vertex = $(-\frac{3}{2}, \frac{25}{2})$,
     y-intercept = $(0, 8)$,
     x-intercepts = $(-4, 0)$ $(1, 0)$

17. y-intercept = $(0, 3)$,
     axis of symmetry: $x = 0$

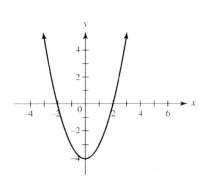

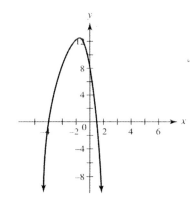

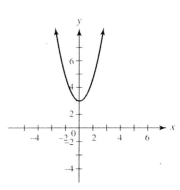

19.

21. y-intercept = $(0, 0)$,
     axis of symmetry: $x = -\frac{1}{6}$

23.

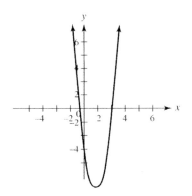

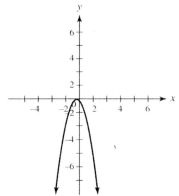

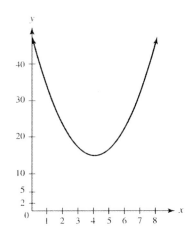

25. 64 feet

Review Exercises Chapter 11

1. $x = -3, x = -4$;
    $(x + 3)(x + 4) = 0$

3. $x = -\frac{1}{2}, x = 6$;
    $(2x + 1)(x - 6) = 0$

5. $n = -\frac{1}{6}, n = 6$;
    $(6n + 1)(n - 6) = 0$

7. $x = \frac{3}{4}$ double root;
    $(4x - 3)(4x - 3) = 0$

9. $R = 0, R = -\frac{7}{5}$;
    $3R(5R + 7) = 0$

11. $t = 10, t = -11$;
     $(t - 10)(t + 11) = 0$

13. $x = -8, x = 3$
   a) $(x+8)(x-3) = 0$
   b) $x = \dfrac{-5 \pm \sqrt{5^2 - 4(1)(-24)}}{2(1)}$

15. $x = -3$ double root;
   a) $(x+3)(x+3) = 0$
   b) $x = \dfrac{-6 \pm \sqrt{6^2 - 4(1)(9)}}{2(1)}$

17. $m = \tfrac{5}{6}, m = 1$;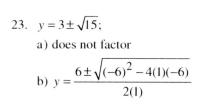
   a) $(6m-5)(m-1) = 0$
   b) $m = \dfrac{11 \pm \sqrt{(-11)^2 - 4(6)(5)}}{2(6)}$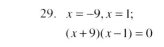

19. $x = 1$ double root;
   a) $3(x-1)(x-1) = 0$
   b) $x = \dfrac{2 \pm \sqrt{(-2)^2 - 4(1)(1)}}{2(1)}$

21. $x = -2 \pm \sqrt{2}$;
   a) does not factor
   b) $x = \dfrac{-4 \pm \sqrt{4^2 - 4(1)(2)}}{2(1)}$

23. $y = 3 \pm \sqrt{15}$;
   a) does not factor
   b) $y = \dfrac{6 \pm \sqrt{(-6)^2 - 4(1)(-6)}}{2(1)}$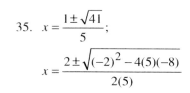

25. $x = 6$ or $x = -5$;
   $x^2 - x = 30$
   $x^2 - x + \dfrac{1}{4} = 30 + \dfrac{1}{4}$
   $\left(x - \dfrac{1}{2}\right)^2 = \dfrac{121}{4}$
   $x - \dfrac{1}{2} = \pm \dfrac{11}{2}, x = \dfrac{1}{2} \pm \dfrac{11}{2}$

27. $x = 1 \pm \sqrt{6}$;
   $x^2 - 2x = 5$
   $x^2 - 2x + 1 = 5 + 1$
   $(x-1)^2 = 6$
   $x - 1 = \pm \sqrt{6}$

29. $x = -9, x = 1$;
   $(x+9)(x-1) = 0$

31. $x = \dfrac{1 \pm \sqrt{11}}{2}$;
   $x = \dfrac{2 \pm \sqrt{(-2)^2 - 4(2)(-5)}}{2(2)}$

33. $x = 1 \pm 2\sqrt{2}$;
   $x^2 - 2x = 7$
   $x^2 - 2x + 1 = 7 + 1$
   $(x-1)^2 = 8$
   $x - 1 = \sqrt{8}$

35. $x = \dfrac{1 \pm \sqrt{41}}{5}$;
   $x = \dfrac{2 \pm \sqrt{(-2)^2 - 4(5)(-8)}}{2(5)}$

37. vertex $= \left(\tfrac{1}{4}, \tfrac{-9}{8}\right)$,
   y-intercept $= (0, -1)$;

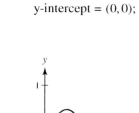
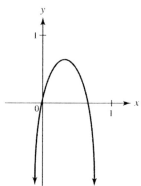

39. vertex $= \left(\tfrac{1}{6}, \tfrac{1}{12}\right)$,
   y-intercept $= (0, 0)$;

41. $x = 1.2, x = -1.7$;
   $x = \dfrac{-1 \pm \sqrt{1^2 - 4(2)(-4)}}{2(2)}$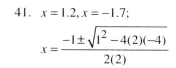

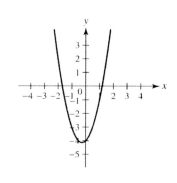

43. no real roots;

$$x = \frac{-1 \pm \sqrt{1^2 - 4(3)(2)}}{2(3)}$$

45. $x(x-14) = 351$

    $x = 27$

47. $2x(x+1) = (x+1)(x+4)$

    4 by 5 inches

49. $3 = 3x^2 - 12x + 13$

    $x = \frac{6 - \sqrt{6}}{3}$

51. $0 = 25x - x^2 - 100$

    $x = 5, x = 20$

53. $x = \frac{-b}{2a} = \frac{-69.7}{2(-1.06)}$

    $x = 32.88$ psi

Chapter 11 Test

1. $x = 2 \pm \sqrt{2}$;

    $x^2 + 4x = -2$

    $x^2 + 4x + 4 = -2 + 4$

    $(x+2)^2 = 2$

    $x + 2 = \pm\sqrt{2}$

2. $x = \frac{1}{2}, x = 4$

    $(2x-1)(x-4) = 0$

3. $x = \frac{2 \pm \sqrt{7}}{2}$;

    $x = \frac{8 \pm \sqrt{(-8)^2 - 4(4)(-3)}}{2(4)}$

4. $x = \frac{6}{5}$;

    $2x^2 - x = 6 - 6x + 2x^2$

    $5x = 6$

5. $x = 3 \pm 3\sqrt{2}$;

    $x^2 - 6x = 9$

    $x^2 - 6x + 9 = 9 + 9$

    $(x-3)^2 = 18$

    $x - 3 = \pm 3\sqrt{2}$

6. opens up,

    vertex = $(-2, -3)$,

    y-intercept = $(0, 5)$;

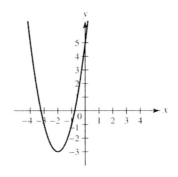

7. $x = 0.8, x = 4.8$;

$$x = \frac{-17 \pm \sqrt{17^2 - 4(-3)(-12)}}{2(-3)}$$

8. $t = 2$ sec;

$$80 = -16t^2 + 22t + 100$$

$$0 = -2(8t^2 - 11t - 10)$$

$$t = \frac{11 \pm \sqrt{(-11)^2 - 4(8)(-10)}}{2(8)}$$

9. $x(x + 6) = 520$

20 by 26 feet

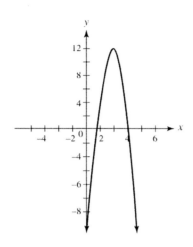

Chapter 12
Exponential and Logarithmic Functions

12.1 Exponential Functions

1. yes

3. yes

5. no, variable must be the exponent, not the base

7. no, the base must be positive

9. 3

11. π

13. increasing, base is > 1

15. increasing, base is > 1

17. $1/16$

19. 18.6

21.

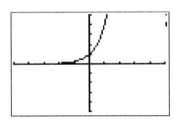

23.

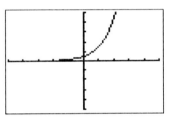

25.

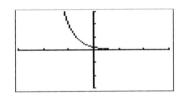

27. $303.88;
$250(1.05)^4$

29. 4.92 mg;
$A(10) = 25(0.85)^{10}$

31. 631.83 million;
$C(t) = 0.0272(1.495)^{2005-1980}$

12.2 Logarithms

1. $\log_{10} 100 = 2$

3. $\log_{10} 0.01 = -2$

5. $\log_{10} 2884 = 3.4600$

7. $\log_{10} 0.0003594 = -3.4444$

9. $\log_2 1024 = 10$

11. $\log_6 216 = 3$

13. $10^1 = 10$

15. $10^3 = 1000$

17. $10^{-2} = 0.01$

19. $10^{2.7536} = 567$

21. $2^3 = 8$

23. $3^5 = 243$

25. 4 since $10{,}000 = 10^4$

27. -3 since $0.001 = 10^{-3}$

29. 6

31. -3

33. 4 since $16 = 2^4$

35. 3 since $64 = 4^3$

37. 0.8609

39. 3.4150

41. 2.3181

43. -2.4461

45. 2 since $4^2 = 16$

47. $\dfrac{-1}{2}$ since $16^{\frac{-1}{2}} = \dfrac{1}{4}$

49. 343 since $7^3 = 343$

51. 256 since $2^8 = 256$

53. 9 since $9^2 = 81$

55. $\dfrac{1}{64}$ since $\left(\dfrac{1}{64}\right)^{\frac{-1}{3}} = 4$

57.

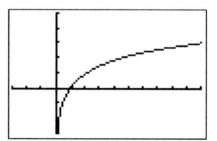

59.

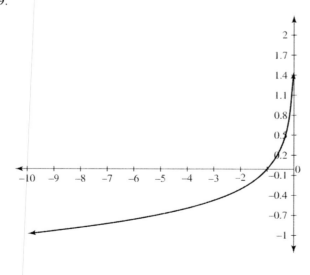

61. 8.40;

$$R = \log\left(\frac{\left(2.5 \times 10^8\right) I_0}{I_0}\right)$$

63. 20 decibels;

10 log 100

12.3 Properties of Logarithms

1. $\log a + \log b$

3. $\log x + \log y + \log z$

5. $\log x - \log 3$

7. $5 \log a$

9. $\log 2 + \log a + \log c - \log 3 - 2\log b$

11. $\frac{1}{2}\log x - 2\log a$

13. $\log (ac)$

15. $\log\left(\frac{9}{3}\right)$

17. $\log\left(x^2 \sqrt{x}\right)$

19. $\log\left[\left(2^2\right) n^3\right]$
   $= \log\left(4n^3\right)$

21. -1.5051;

   $\log 1 - \log 32$

23. 1.1451;

   2.4(log 3)

25. 0.3578;

   $\frac{1}{4}(\log 27)$

27. 1.8751;

   $\log 3 + \log 25$

29. 6.4;

   $\text{pH} = -\log(3.97 \times 10^{-7})$

31. 20;

   $S = 10 \log\left(\frac{100 I_0}{I_0}\right)$

33. 1.6812;

   $n = 10 \log\left(\frac{12}{0.25}\right)$

35. 6556.66;

   $E = (85)(265) \log\left(\frac{518}{265}\right)$

37. 4.56×10^{192};

   $x = 16^{160}$
   $\log x = 160 \log 16$
   $\log x = 192.659$
   $10^{192.659} = x$
   $x = 10^{0.659} \times 10^{192}$

39. 3.27×10^{150};

   $x = 2^{500}$
   $\log x = 500 \log 2$
   $\log x = 150.515$
   $10^{150.515} = x$
   $x = 10^{0.515} \times 10^{150}$

41. 1.34×10^{154};

$x = 16^{128}$

$\log x = 128 \log 16$

$\log x = 154.127$

$10^{154.127} = x$

$x = 10^{0.127} \times 10^{154}$

43. 3.76×10^{414};

$x = 16^{256} \times 455^{40}$

$\log x = 256 \log 16 + 40 \log 455$

$\log x = 414.575$

$10^{414.575} = x$

$x = 10^{0.575} \times 10^{414}$

12.4 Natural Logarithms

1. 2.1281

3. −0.1054

5. 5.2983

7. −6.9078

9. $x = 9.8956$;

$x = (6.92)(1.43)$

$\ln x = \ln(6.92) + \ln(1.43)$

$\ln x = 2.292$

$e^{2.292} = x$

11. $x = 304.226$;

$\ln x = \ln(81,350) - \ln(267.4)$

$\ln x = 5.7178$

$e^{5.7178} = x$

13. $x = 28.0535$;

$\ln x = \dfrac{1}{2}\ln(787$

$\ln x = 3.3341$

$e^{3.3341} = x$

15. $x = 5.46 \times 10^{18}$;

$\ln x = 50\ln(2.37)$

$\ln x = 43.1445$

$e^{43.1445} = x$

17.   2.708;

$\ln(15) = \ln(3) + \ln(5)$

$= 1.0986 + 1.6094$

19.   4.8282;

$\ln(125) = 3\ln(5)$

$= 3(1.6094)$

21. 4.46;

$t = \dfrac{-\ln\left(\dfrac{400}{500}\right)}{0.05}$

23. 0.099;

$i = \dfrac{\ln(2)}{7}$

25. 0.085;

$r = \dfrac{\ln\left(\dfrac{645.23}{500}\right)}{3}$

27. 2.159;

$r = \dfrac{\ln\left(\dfrac{0.00000015}{0.000000002}\right)}{2}$

29. 2.545;

$C = \dfrac{4.92}{(\ln 22)^{0.584}}$

12.5 Exponential and Logarithm Equations

1. $x = 4$;

$2^x = 2^4$

3. $x = -0.7481$;

$5^x = 0.3$

$x \log 5 = \log 0.3$

$x = \dfrac{\log 0.3}{\log 5}$

5. $x = -2.8645$;

$e^{-x} = 17.54$

$-x = \ln 17.54$

7. $x = 1.4307;$

$$(x-1)\log 5 = \log 2$$

$$x - 1 = \frac{\log 2}{\log 5}$$

9. $x = -0.7570;$

$$10^{x+2} = \frac{35}{2}$$

$$x + 2 = \log\left(\frac{35}{2}\right)$$

$$x = \log\left(\frac{35}{2}\right) - 2$$

11. $x = 1.9459;$

$$e^x = 7$$

$$x = \ln 7$$

13. $x = 2.7093$

15. $x = \frac{1}{8};$

$$\log_{32} x = \frac{-3}{5}$$

$$32^{-3/5} = x$$

$$2^{-3} = x$$

17. $x = 1.5772;$

$$\log(2x-1) = \frac{1}{3}$$

$$\sqrt[3]{10} = 2x - 1$$

$$\frac{\sqrt[3]{10} + 1}{2} = x$$

19. $x = 250;$

$$\log\left(\frac{12x^2}{3x}\right) = 3$$

$$\log(4x) = 3$$

$$10^3 = 4x$$

21. $x = 64.5;$

$$\ln\left(\frac{2x-1}{4^2}\right) = \ln 2^3$$

$$\frac{2x-1}{16} = 8$$

$$2x - 1 = 128$$

$$2x = 129$$

23. $x = \dfrac{-7 + \sqrt{161}}{4};$

$$\log(2x-1)(x+4) = 1$$

$$10 = (2x-1)(x+4)$$

$$10 = 2x^2 + 8x - x - 4$$

$$0 = 2x^2 + 7x - 14$$

$$x = \frac{-7 \pm \sqrt{7^2 - 4(2)(-14)}}{2(2)}$$

25. $t = 1.6778;$

$$35 = 22 + 98(0.30)^t$$

$$13 = 98(0.30)^t$$

$$\frac{13}{98} = (0.30)^t$$

$$\log\left(\frac{13}{98}\right) = t\log(0.30)$$

$$\frac{\log\left(\frac{13}{98}\right)}{\log(0.30)} = t$$

27. $h = 104.830;$

$$130 = 64e^{0.00676h}$$

$$\ln\left(\frac{130}{64}\right) = 0.00676h$$

29. $r = 31.1967\%;$

$$2600 = 400e^{r \times 6}$$

$$\frac{2600}{400} = e^{6r}$$

$$\ln\left(\frac{2600}{400}\right) = 6r$$

12.6 Log-Log and Semi-Log Graphs

1.

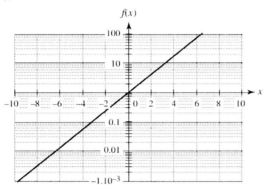

3.

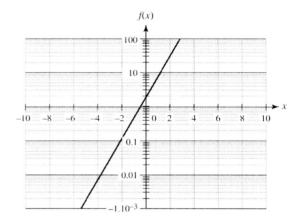

5.

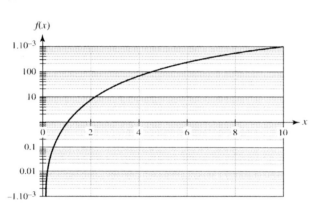

7.

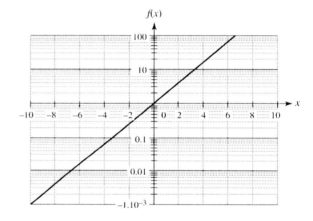

9.

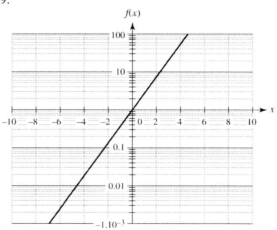

11.

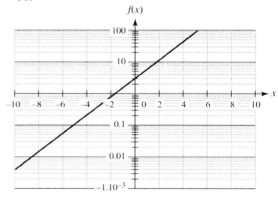

13.

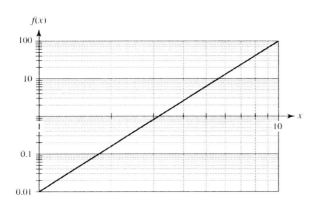

15.

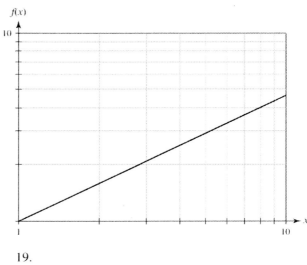

17.

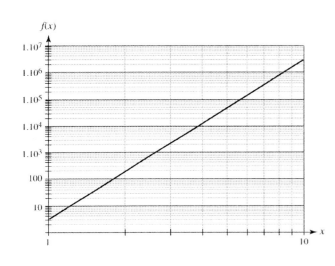

19.

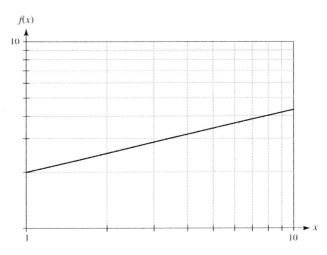

21.

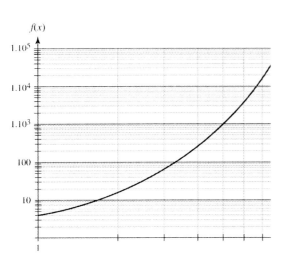

23.

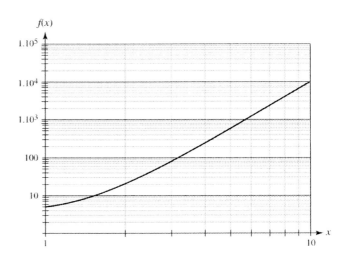

25.

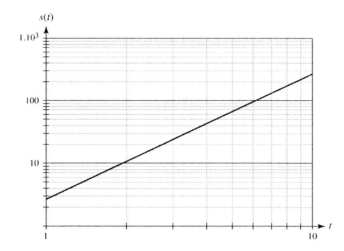

27a.   27b.

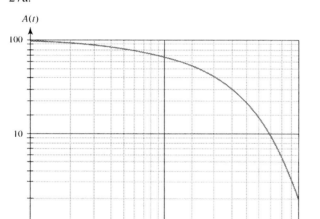   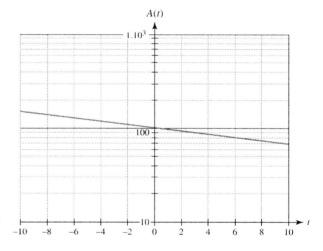

Review Exercises for Chapter 12

1. $\log_{10} 10,000 = 4$

3. $\log_6 1296 = 4$

5. $\log_{10} 10 = 1$

7. $\log_{10} 0.00288 = -2.54$

9. $2^7 = 128$

11. $10^5 = 100,000$

13. $e^{3.0} = 20$

15. $4^5 = 1024$

17. $\log(2.07) + \log(3.45)$

19. $\log x + \log y$

21. $\log 2 + \log a + \log b + \log c$

23. $\log(89.15) - \log(9.176)$

25. $\log a + \log b - \log c - \log d$

27. $0.3\log(1.034)$

29. $5\log a + 2\log b$

31. $\dfrac{1}{3}\log(0.9006)$

33. $\dfrac{1}{2}\log(86,000) - \log(45.8)$

35. $\dfrac{1}{2}\log 2 + \dfrac{1}{2}\log x - 2\log a$

37. 1.4110

39. -0.5108

41. $1.7917;$
$\quad \ln 2 + \ln 3 = 0.6931 + 1.0986$

43. $0.4055;$
$\quad \ln 3 - \ln 2 = 1.0986 - 0.6931$

45. $\log(4c)$

47. $\log\left(\sqrt{\dfrac{7}{x}}\right)$

49. $\log[(\sqrt{2x})^{-1}(y^2)]$

51. $\log[(2x)^2(y^3)]$

53. $\log\left(\dfrac{y}{x^2}\right)^2$

55. $\log\left(\dfrac{y}{x^3}\right)^2$

57. $x = 0.8047;$
$\quad e^{2x} = 5$
$\quad 2x = \ln 5$
$\quad x = \dfrac{\ln 5}{2}$

59. $\qquad x = 4.3013;$
$\qquad 3^{x+2} = 5^x$
$\qquad (x+2)\log 3 = x\log 5$
$\qquad x\log 3 + 2\log 3 = x\log 5$
$\qquad\quad 2\log 3 = x(\log 5 - \log 3)$
$\qquad\quad \dfrac{2\log 3}{\log 5 - \log 3} = x$

61. $\qquad x = 2;$
$\quad \log x + \log 6 = \log 12$
$\qquad \log x = \log 12 - \log 6$
$\qquad\quad \log x = \log\left(\dfrac{12}{6}\right)$
$\qquad\quad \log x = \log 2$

63. $\qquad x = 48;$
$\quad \log(x+2) = 2 - \log 2$
$\quad \log(x+2) + \log 2 = 2$
$\qquad \log(2x+4) = 2$
$\qquad\quad 10^2 = 2x + 4$
$\qquad\quad 96 = 2x$

65.   67.

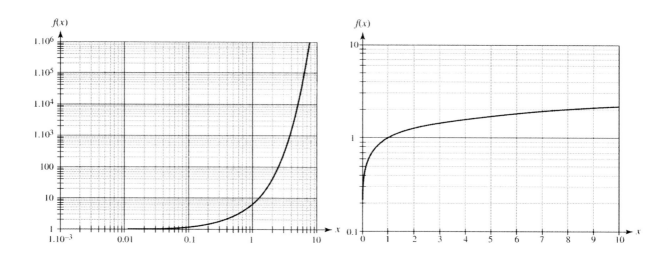

69. $m = 4.64$;

$\quad m = 3.32\log_{10}(25)$

71. 140 decibels;

$\quad 10\log\dfrac{1\times10^{14}I_0}{I_0}$

$\quad 10\log(10^{14})$

$\quad 10(14)$

73. $H^+ = 0.0003922$;

$\quad 3.4065 = -\log(H^+)$

$\quad -3.4065 = \log(H^+)$

$\quad 10^{-3.4065} = H^+$

75. 15.8;

$\quad T = 50.0(10^{-0.1(5)})$

77. $P = 6.1553$;

$\quad \ln P = 3.00 - 1.50\ln(2.20)$

$\quad \ln P = 1.8173$

$\quad e^{1.8173} = P$

Chapter 12 Test

1. -0.0102

2. 2.0491

3. $x = 1.4307$;

$\quad 5^{x-1} = 2$

$\quad (x-1)\log 5 = \log 2$

$\quad x - 1 = \dfrac{\log 2}{\log 5}$

$\quad x = \dfrac{\log 2}{\log 5} + 1$

4. $x = 14.1530;$

$$15.6^{x+2} = 23^x$$
$$(x+2)\log 15.6 = x\log 23$$
$$x\log 15.6 + 2\log 15.6 = x\log 23$$
$$2\log 15.6 = x(\log 23 - \log 15.6)$$
$$\frac{2\log 15.6}{\log 23 - \log 15.6} = x$$

5. $x = -0.1623;$

$$\log(3-x) = \frac{1}{2}$$
$$10^{1/2} = 3 - x$$
$$x = 3 - \sqrt{10}$$

6. $x = 0.9061;$

$$\ln x - \ln\frac{1}{3} = 1$$
$$\ln 3x = 1$$
$$e^1 = 3x$$
$$x = \frac{e}{3}$$

7.

8. $\log 4 + 5\log a - \log 9$

9. $t = 8.66$ years;

$$2A_0 = A_0 e^{0.08t}$$
$$2 = e^{0.08t}$$
$$\ln 2 = 0.08t$$
$$t = \frac{\ln 2}{0.08}$$

Chapter 13
Geometry and Right Triangle Trigonometry

13.1 Angles and Their Measure

1. $53°$

3. $21°$

5. $\angle CEB, \angle CED$

7. $\angle DEC$ and $\angle CEB$, $\angle CEB$ and $\angle BEA$

9. $\angle ABE, \angle CBE$

11. $\angle CBD$

13. $25°; 90 - 65$

15. $115°; 90 + 25$

17. $\angle 1 = 40°, 180 - 140;$ $\angle 2 = 40°; \angle 4 = 40°$

19. $\angle BOC = 62°, 90 - 28;$ $\angle EOF = 62°, \angle EOF = \angle BOC$

21. 1 and 5, 3 and 4

23. 1 and 3, 4 and 5

25. $\angle 1 = 50°;$ corresponding angles

27. $\angle 4 = 130°;$ alternate exterior and corresponding angles

29. $58°; 148 - 90$

31. $148°;$ corresponding angles

33. $40°; 90 - 50$

35. $50°;$ corresponding angles

37. $40°; 180 - 100 - 40$

39. $100°;$ supplementary angles

13.2 Other Geometric Figures

1. $56°; 180 - 84 - 40$

3. $48°; 180 - 66 - 66$

5. $68°; 360 - 95 - 155 - 42$

7. $80°; \dfrac{360 - 100 - 100}{2}$

9. $60°; 180 - 90 - 30$

11. $120°; 180 - 60$

13. $32°;$ central angle

15. $76°;$ inscribed angle

17. DG, BC

19. BC, CG

21. $\angle BOG$

23. BG, CG

25. $60°;$ central angle

27. $30°;$ inscribed angle

29. $110°;$ inscribed angle

31. $35°; 90 - 55$

33. three

35. five

37. $40°$

39. $135°;$
   total degrees$=180(n - 2)$
   $180(8 - 2) = 180(6) = 1080$
   $\dfrac{1080}{8} = 135$

41. $117°;$
   $180 - 50 - 67 = 63,$
   $180 - 63 = 117$

13.3 Right Triangles and Pythagorean Theorem

1. $5; \sqrt{3^2 + 4^2}$

3. $17; \sqrt{8^2 + 15^2}$

5. $8; \sqrt{10^2 - 6^2}$

7. $4.90; \sqrt{7^2 - 5^2}$

9. $10.6; \sqrt{16^2 - 12^2}$

11. $28.3; \sqrt{32^2 - 15^2}$

13. $59.9; \sqrt{82^2 - 56^2}$

15. $40.9; \sqrt{5.62^2 + 40.5^2}$

17. $2.67; \sqrt{2.76^2 - 0.709^2}$

19. $39.1; \sqrt{42.4^2 - 16.5^2}$

21. $5.66; \sqrt{4.00^2 + 4.00^2}$

23. $\quad a = 12.02;$
   $$a^2 + a^2 = 17^2$$
   $$2a^2 = 289$$
   $$a^2 = 144.5$$
   $$a = \sqrt{144.5}$$

25. 7.89 m;
   $\sqrt{7.8^2 + 1.2^2}$

27. 521.39 ft;
   $\sqrt{520^2 + 38^2}$

29. 36.77 ft;
   $\sqrt{19.5^2 - 6.5^2} = 18.38,$
   $18.38 \times 2 = 36.77$

31. $z = 18.90;$
   $\sqrt{17.0^2 + 8.25^2}$

33. 893.78 lb;
   $\sqrt{865^2 + 225^2}$

35. 8.42 m;
   $\sqrt{4.75^2 + 6.50^2} = 8.05,$
   $\sqrt{8.05^2 + 2.45^2} = 8.42$

37. 161.78 ft;
   $\sqrt{81.3^2 + 92.4^2} = 123.07$
   $\sqrt{123.07^2 + 105^2} = 161.78$

39. 56.20 cm;
   $x^2 + 4x = 576$
   $5x^2 = 576$
   $x^2 = 115.2$
   $x = 10.73$

13.4 Similar Triangles

1. $\angle A' = 20°, \angle B' = 100°, \angle C' = 60°$

3. $B'C' = 16.5, A'C' = 19.5;$
$$\frac{10}{15} = \frac{11}{B'C'}, \ \frac{10}{15} = \frac{13}{A'C'}$$

5. $\angle B' = 65°, \angle C' = 75°,$
$\angle A' = 180 - 65 - 75 = 40°$

7. $B'C' = 12.7;$
$\angle A' = \angle B' = 70°;$
$$\frac{180 - 40}{2} = 70$$

9. similar

11. neither

13. $\angle E, EF$

15. $\angle U, TS$

17. $DE = 10;$
$$\frac{8}{4} = \frac{DE}{5}$$

19. $TU = 12;$
$$\frac{9}{6} = \frac{TU}{8}$$

21. $CE = 21;$
$$\frac{18}{6} = \frac{CE}{7}$$

23. $BA = 20;$
$$\frac{18}{6} = \frac{CA}{10}$$
$CA = 30$

25. $AB = 16;$
$$\frac{9}{12} = \frac{12}{AB}$$

27. $LM = 8;$
$$\frac{9}{12} = \frac{6}{LM}$$

29.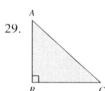

31. $x = \frac{10}{3}, y = 4;$
$$\frac{3}{2} = \frac{5}{x} = \frac{6}{y}$$

33. $\angle K = \angle K$ (identity),
$\angle Y = \angle F$ (alternate interior $\angle$s),
$\angle X = \angle N$ (alternate interior $\angle$s)

35. $8; \frac{16}{2}$

37. $x = 6\frac{2}{3}$ ft;
$$\frac{4}{3} = \frac{x}{5}$$

39. $x = 3.5$ m;
$$\frac{1}{80} = \frac{x}{280}$$

41. $x = 2.25$ in;
$$\frac{10}{1\frac{1}{4}} = \frac{18}{x}$$

43. $x = 1525.68$ km, $y = 1804.41$ km,
$z = 762.84$ km;
$$\frac{1}{146.7} = \frac{10.4}{x} = \frac{12.3}{y} = \frac{5.2}{z}$$

45. $x = 5\frac{1}{3}$ ft;
$$\frac{6}{4} = \frac{8}{x}$$

13.5 The Trigonometric Ratios

1. $\sin A = \dfrac{9}{41}, \tan A = \dfrac{9}{40},$

$\cos B = \dfrac{9}{41}$

3. $\cot B = \dfrac{9}{40}, \sec A = \dfrac{41}{40},$

$\tan B = \dfrac{40}{9}$

5. $\sin A = \dfrac{1}{2}, \sec B = 2,$

$\cot A = \sqrt{3}$

7. $\tan A = \dfrac{1}{\sqrt{3}}, \cos B = \dfrac{1}{2},$

$\sec A = \dfrac{2}{\sqrt{3}}$

9. $\cos A = \dfrac{2.47}{3.96} = 0.62$

$\tan A = \dfrac{3.10}{2.47} = 1.26$

$\csc B = \dfrac{3.96}{2.47} = 1.60$

11. $\sin A = \dfrac{3.10}{3.96} = 0.78$

$\cos B = \dfrac{3.10}{3.96} = 0.78$

$\cot A = \dfrac{2.47}{3.10} = 0.80$

13. $\sin A = \dfrac{4}{5}, \tan B = \dfrac{3}{4};$

$c = \sqrt{4^2 + 3^2} = 5$

15. $\cot A = \dfrac{7}{24}, \cos B = \dfrac{24}{25};$

$a = \sqrt{25^2 - 7^2} = 24$

17. $\cos A = \dfrac{1}{\sqrt{2}}, \cot B = 1;$

$c = \sqrt{1^2 + 1^2} = \sqrt{2}$

19. $\csc A = \dfrac{22}{\sqrt{259}}, \cos B = \dfrac{\sqrt{259}}{22};$

$a = \sqrt{22^2 - 15^2} = \sqrt{259}$

21. $\sin A = \dfrac{1.2}{1.92} = 0.63,$

$\tan B = \dfrac{1.5}{1.2} = 1.25;$

$c = \sqrt{1.2^2 + 1.5^2} = 1.92$

23. $\sin A = \dfrac{0.0413}{0.0608} = 0.679,$

$\sec B = \dfrac{0.0608}{0.0413} = 1.472;$

$a = \sqrt{0.0608^2 - 0.0446^2}$

$= 0.0413$

25. $\sin A = \dfrac{1}{\sqrt{2}};$

$a = 1, b = 1 \Rightarrow$

$c = \sqrt{1^2 + 1^2} = \sqrt{2}$

27. $\csc A = 1.40;$

$b = 0.70, c = 1 \Rightarrow$

$a = \sqrt{1^2 - 0.70^2} = 0.71$

$\Rightarrow \csc A = \dfrac{1}{0.71}$

29. $\cos A = 0.49;$

$b = 0.563, a = 1 \Rightarrow$

$c = \sqrt{1^2 + 0.563^2} = 1.15$

$\Rightarrow \cos A = \dfrac{0.563}{1.15}$

31. $\tan A = 1.04;$

$a = 0.720, c = 1 \Rightarrow$

$b = \sqrt{1^2 - 0.720^2} = 0.69$

$\Rightarrow \tan A = \dfrac{0.720}{0.69}$

33. $\sin A = 0.5;$

$b = 0.8660, c = 1 \Rightarrow$

$a = \sqrt{1^2 - 0.8660^2} = 0.5$

$\Rightarrow \sin A = \dfrac{0.5}{1}$

35. $\tan A = \sqrt{2};$

$c = \sqrt{3.00}, b = 1 \Rightarrow$

$a = \sqrt{\sqrt{3}^2 - 1^2} = \sqrt{2}$

$\Rightarrow \tan A = \dfrac{\sqrt{2}}{1}$

37. $\tan A = \dfrac{a}{b} = \cot B$

$\csc A = \dfrac{c}{a} = \sec B$

39. $\sin A = \dfrac{a}{c}$ $\csc A = \dfrac{c}{a}$

$\cos A = \dfrac{b}{c}$ $\sec A = \dfrac{c}{b}$

$\tan A = \dfrac{a}{b}$ $\cot A = \dfrac{b}{a}$

41. $c = 13$,

$\left(\dfrac{5}{13}\right)^2 + \left(\dfrac{12}{13}\right)^2 = 1$

43. $c = 13$;

$\sin A = \dfrac{5}{13}, \cos A = \dfrac{12}{13},$

$\tan A = \dfrac{5}{12}$;

$\dfrac{\sin A}{\cos A} = \dfrac{5/13}{12/13} = \dfrac{5}{12} = \tan A$

13.6 Values of the Trigonometric Ratios

1.   0.8480

3.   0.4557

5.   1.5003

7.   0.9178

9.   0.2382

11.   0.7046

13.   4.2223

15.   2.4751

17.   32.0°

19.   18.9°

21.   60.5°

23.   61.9°

25.   52.7°

27.   18.0°

29.   84.3°

31.   58.2°

33.   1.2062;

$\sin^{-1}(0.5592),$

$1/\cos(34.0°)$

35.   1.1792;

$\tan^{-1}(1.600),$

$1/\sin(58.0°)$

37.   53.1°;

$\sin \alpha = \dfrac{4}{5}$

39.   40.9°;

$\cos \alpha = \dfrac{6.2}{8.2}$

41.   34.6°;

$\sin \alpha = \dfrac{15.5}{27.3}$

43.   79.4°;

$\tan \alpha = \dfrac{6580}{1230}$

45.   13.37 V;

$V_{inst} = 15(\sin 63°)$

47.   126.51 m;

$H = 132.8(\cos 17.7°)$

49.   1.56;

$n = \dfrac{\sin 72.0°}{\sin 37.7°}$

51.   23.9 m;

$(25)(\cos 17.0°)$

13.7 Right Triangle Applications

1. $B = 60°; 90 - 30$

   $c = 24; \sin 30° = \dfrac{12}{c}$

   $b = 20.8; \tan 30° = \dfrac{12}{b}$

3. $A = 33.7°; 90 - 56.3$

   $b = 18.7; \sin 56.3° = \dfrac{b}{22.5}$

   $a = 12.5; \cos 56.3° = \dfrac{a}{22.5}$

5. $B = 13.2°; 90 - 76.8$

   $a = 13.0; \sin 76.8° = \dfrac{a}{13.4}$

   $b = 3.1; \cos 76.8° = \dfrac{b}{13.4}$

7. $A = 28.8°, B = 61.2°,$

   $b = 1.18;$

   $\sin A = \dfrac{0.650}{1.35}, \cos B = \dfrac{0.650}{1.35},$

   $b = \sqrt{1.35^2 - 0.650^2}$

9. $A = 82.6°, B = 7.4°,$

   $a = 44.6;$

   $\cos A = \dfrac{5.80}{45.0}, \sin B = \dfrac{5.80}{45.0},$

   $a = \sqrt{45.0^2 - 5.80^2}$

11. $A = 64.2°, B = 25.8°,$

    $b = 4.71;$

    $\sin A = \dfrac{9.72}{10.8}, \cos B = \dfrac{9.72}{10.8},$

    $b = \sqrt{10.8^2 - 9.72^2}$

13. $a = 1.88, c = 15.41,$

    $B = 83°;$

    $\tan 7.0° = \dfrac{a}{15.3}, \cos 7.0° = \dfrac{15.3}{c},$

    $B = 90 - 7$

15. $A = 33.5°, B = 56.5°,$

    $c = 117.9;$

    $\tan A = \dfrac{65.1}{98.3}, \tan B = \dfrac{98.3}{65.1},$

    $c = \sqrt{65.1^2 + 98.3^2}$

17. $x = 29.1$ ft;

    $\tan 10.0° = \dfrac{x}{165}$

19. $x = 303.5$ cm;

    $\sin 23.7° = \dfrac{122}{x}$

21. $x = 318.0$ cm;

    $\tan 23.0° = \dfrac{135}{x}$

23. $x = 85.2°;$

    $\cos x° = \dfrac{4.25}{50.5}$

25. $x = 4005.7$ ft;

    $\tan 38.7° = \dfrac{x}{5000}$

27. $x = 6.8°;$

    $\tan x° = \dfrac{1.5}{12.5}$

29. $x = 1640.2$ ft;

    $\tan 26.0° = \dfrac{800}{x}$

31. $x = 242.5$ km;

    $\tan 5.3° = \dfrac{22.5}{x}$

33. $x = 4707.2$ ft;

    $\tan 28.68° = \dfrac{2575}{x}$

35. $x = 22.9°;$

    $\tan x° = \dfrac{7350}{17,400}$

Review Exercises Chapter 13

1. $61°; 90 - 29$

3. $90°$

5. $27°; 180 - 90 - 63$

7. $\angle EBC$

9. $\angle EBD$

11. $\angle 2$ and $\angle 5$

13. $180°$

15. $65°; 90 - 25$

17. $32°; 90 - 58$,
    vertical angles

19. $52°; 90 - 38$,
    alternate interior $\angle$s

21. $132°; 90 - 42 = 48$,
    $180 - 48 = 132$

23. $36°; 180 - 72 - 72$

25. $50°$;
    $\angle BCA = 65°$ by isosceles Δ,
    $\angle BAC = 180 - 65 - 65 = 50°$,
    $\angle DCA = 50°$ by alt int $\angle$s

27. $120°$; inscribed angle
    is half the arc

29. $40°$; inscribed angle
    is half the arc

31. $25°$; inscribed angle
    is half the arc

33. $65°; 90 - 25$

35. $41; \sqrt{9^2 + 40^2}$

37. $42; \sqrt{58^2 - 40^2}$

39. $7.36; \sqrt{6.30^2 + 3.80^2}$

41. $21.1; \sqrt{36.1^2 - 29.3^2}$

43. $AB = 7.5$;
    $\dfrac{12}{8} = \dfrac{AB}{5}$

45. 0.735

47. 0.130

49. 0.116

51. 1.16

53. $30.0°$

55. $45.0°$

57. $3.00°$

59. $56.7°$

61. $\sin A = \dfrac{23}{41}, \cos B = \dfrac{23}{41}$

63. $\sec B = \dfrac{41}{23}, \tan A = \dfrac{23}{34}$

65. $\cos A = \dfrac{5}{13}, \tan B = \dfrac{5}{12}$;
    $c = \sqrt{12^2 + 5^2} = 13$

67. $\tan A = \dfrac{8.70}{2.30} = 3.78$,
    $\cos B = \dfrac{2.30}{9.00} = 0.256$;
    $c = \sqrt{8.70^2 + 2.30^2} = 9.00$

69. $\cot A = \dfrac{0.0410}{0.0890} = 0.461$,
    $\sin B = \dfrac{0.0410}{0.0980} = 0.419$;
    $b = \sqrt{0.0980^2 - 0.0890^2}$
    $= \sqrt{0.00168} = 0.0410$

71. $\cos A = \dfrac{0.0943}{0.105} = 0.898$,
    $\tan B = \dfrac{0.0943}{0.0462} = 2.04$;
    $a = \sqrt{0.105^2 - 0.0943^2}$
    $= 0.0462$

73. $A = 38.0°$;
    $\sin A = \dfrac{8.00}{13.0}$

75. $B = 13.1°$;
    $\tan B = \dfrac{1.30}{5.60}$

77. $B = 28.4°$;
    $\sin B = \dfrac{3420}{7200}$

79. $A = 53.4°$;

$$\tan A = \frac{4910}{3650}$$

81. $a = 2.48, b = 6.47,$
$B = 69.0°$;

$$\sin 21° = \frac{a}{6.93}, \cos 21° = \frac{b}{6.93},$$

$$B = 90 - 21$$

83. $c = 85.0, a = 71.5,$
$A = 57.3°$;

$$\sin 32.7° = \frac{45.9}{c},$$

$$\tan 32.7° = \frac{45.9}{a},$$

$$A = 90 - 32.7$$

85. $A = 59.1°, B = 30.9°,$
$c = 10.1$;

$$\tan A = \frac{8.70}{5.20}, \tan B = \frac{5.20}{8.70},$$

$$c = \sqrt{8.70^2 + 5.20^2}$$

87. $A = 63.9°, B = 26.1°,$
$b = 47.5$;

$$\sin A = \frac{97.0}{108}, \cos B = \frac{97.0}{108},$$

$$b = \sqrt{108^2 - 97.0^2}$$

89. $a = 8396.6, b = 11,042.1,$
$B = 52.75°$;

$$\sin 37.25° = \frac{a}{13,872},$$

$$\cos 37.25° = \frac{b}{13,872},$$

$$B = 90 - 37.25$$

91. $A = 88.9°, B = 1.1°,$
$a = 112.483$;

$$\cos A = \frac{2.13520}{112.503},$$

$$\sin B = \frac{2.13520}{112.503},$$

$$a = \sqrt{112.503^2 - 2.13520^2}$$

93. 21.1 ft;

$$\sqrt{12.5^2 + 17.0^2}$$

95. 56.1 m;

$$24^2 = x^2 + (2x)^2$$

$$576 = 5x^2$$

$$115.2 = x^2$$

$$10.7 = x$$

$$56.1 = 24 + 10.7 + 2(10.7)$$

97. $22.5, 67.5$;

$$x + 3x = 90$$

$$4x = 90$$

$$x = 22.5$$

99. 5 ft;

$$\sqrt{13^2 - 12^2}$$

101. $x = 5.03$ cm;

$$\sqrt{8.25^2 - 6.38^2} = 5.23,$$

$$\frac{5.23}{6.38} = \frac{4.12}{x}$$

103. $x = 1.5$ m;

$$\frac{8.00}{2.50} = \frac{4.80}{x}$$

105. 55.1 mA;

$$i = 56.0(\cos 10.5°)$$

107. 1211.8 ft-lb;

$$W = (57.3)(23.8)(\cos 27.3°)$$

109. $x = 77.5$ ft;

$$\tan 40.0° = \frac{65.0}{x}$$

111. $\theta = 63.4°, 26.6°$;

$$\tan \theta = \frac{8}{4}$$

113. $x = 28.1°$;

$$\tan x = \frac{8}{15}$$

115. $x = 6433.5$ ft;

$$\tan 65.0° = \frac{x}{3000}$$

117. $x = 16.1°$;

$$\tan x = \frac{150}{520}$$

119. $x = 2027$ mi;

$$\sin 1.8° = \frac{2}{c}, c = 63.7,$$

$$\tan 1.8° = \frac{63.7}{x}$$

121. no, a counterexample would be

$\cos 30° = 0.8660$ but $\cos 60° = 0.5$;

yes, proof:

$A > B$ implies $a > b$.

Thus, $\dfrac{a}{c} > \dfrac{b}{c}$ and

$\cos B > \cos A$

Chapter 13 Test

1. $\angle CBD, \angle DBE$

2. $\angle ABD$

3. $\angle DBE = 25°$

4. $\angle ABD = 115°$

5. $\angle DBE$

6. $62°$; vertical angles

7. $118°$; alternate interior $\angle$s, supplemental $\angle$s

8. $28°$; complementary $\angle$s, alternate interior $\angle$s

9. $152°$; supplementary $\angle$s

10. $LM = 8$;

$\dfrac{9}{12} = \dfrac{6}{LM}$

11. a) 0.909

b) $49.0°$

12. $\tan A = \dfrac{2}{\sqrt{5}}$

13. $\cot A = 0.7728$

14. $c = 66.1, a = 39.8,$

$B = 53°$;

$\cos 37° = \dfrac{52.8}{c}$,

$\tan 37° = \dfrac{a}{52.8}$,

$B = 90 - 37$

15. $x = 67.0$ ft;

$\sin 52° = \dfrac{x}{85}$

Chapter 14
Oblique Triangles and Vectors

14.1 Trigonometric Functions of Any Angle

1.

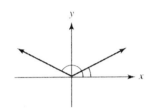

3.

5. $r = 13 = \sqrt{5^2 + 12^2}$,

$$\sin\theta = \frac{12}{13}, \tan\theta = \frac{12}{5}$$

7. $r = \sqrt{13} = \sqrt{(-3)^2 + 2^2}$,

$$\sec\theta = \frac{\sqrt{13}}{-3}, \sin\theta = \frac{2}{\sqrt{13}}$$

9. $r = \sqrt{2} = \sqrt{(-1)^2 + (-1)^2}$,

$$\cot\theta = \frac{-1}{-1} = 1, \cos\theta = \frac{-1}{\sqrt{2}}$$

11. $r = 10 = \sqrt{6^2 + (-8)^2}$,

$$\cos\theta = \frac{6}{10} = \frac{3}{5},$$

$$\sin\theta = \frac{-8}{10} = \frac{-4}{5}$$

13. $r = \sqrt{10} = \sqrt{1^2 + 3^2}$,

$$\cos\theta = \frac{1}{\sqrt{10}} = 0.32,$$

$$\cot\theta = \frac{1}{3} = 0.33$$

15. $r = 16.23$

$$= \sqrt{(-15.0)^2 + 6.20^2},$$

$$\tan\theta = \frac{6.20}{-15.0} = -0.41,$$

$$\csc\theta = \frac{16.23}{6.20} = 2.62$$

17. $r = 220.23$

$$= \sqrt{(-140)^2 + (-170)^2},$$

$$\sin\theta = \frac{-170}{220.23} = -0.77,$$

$$\sec\theta = \frac{220.23}{-140} = -1.57$$

19. $r = 32.4 = \sqrt{27.3^2 + (-17.5)^2}$,

$$\csc\theta = \frac{32.4}{-17.4} = -1.86,$$

$$\cos\theta = \frac{27.3}{32.4} = 0.84$$

21. $+, -, -$

23. $-, -, -$

25. $+, +, +$

27. $-, -, -$

29. IV

31. I

33. II

35. II

37. $\sin = +, \cos = - \Rightarrow$ II

39. $\sin = +, \cos = - \Rightarrow$ II

41. $\sin = +, \cos = - \Rightarrow$ II

43. $\sin = 1, \cos = 0 \Rightarrow 90°$

45. $180 - 165 = 15, \sin 15°$;
$230 - 180 = 50, -\cos 50°$

47. $207 - 180 = 27, -\cos 27°$;
$360 - 290 = 70, -\csc 70°$

49. $360 - 342 = 18, -\tan 18°$;
$\sec 10°$

51. $650 - 360 = 290$,
$360 - 290 = 70, -\cot 70°$;
$360 - 300 = 60, \tan 60°$

53. -0.5446

55. -10.7797

57. -1.0904

59. -0.9265

61. 3.7321

63. -1.7179

65. 0.9759

67. 2.7776

69. -0.5358

71. 2.4142

73. $238.0°, 302.0°$;
$\phi = 58.0°$,
$180 + 58.0, 360 - 58.0$

75. $66.4°, 293.6°$;
$\phi = 66.4°$,
$0 + 66.4, 360 - 66.4$

77. $62.2°, 242.2°$;
$\phi = 62.2°$,
$0 + 62.2, 180 + 62.2$

79. $15.8°, 195.8°$;
$\phi = 15.8°$,
$0 + 15.8, 180 + 15.8$

81. $232.0°, 308.0°$;
$\phi = 52.0°$,
$180 + 52.0, 360 - 52.0$

83. $219.3°, 320.7°$;
$\phi = 39.3°$,
$180 + 39.3, 360 - 39.3$

85. $167.8°, 192.2°$;
$\phi = 12.2°$,
$180 - 12.2, 180 + 12.2$

87. $178.0°, 182.0°$;
$\phi = 2.0°$,
$180 - 2.0, 180 + 2.0$

89. $334.0°$;
$\phi = 26.0°, 360 - 26.0$

91. $129.8°$;
$\phi = 50.2°, 180 - 50.2$

93. $119.6°$;
$\phi = 60.4°, 180 - 60.4$

95. $189.2°$;
$\phi = 9.2°, 180 + 9.2$

97. $\theta = 85.2°$
$= \tan^{-1}\left(\dfrac{1250}{105}\right)$

99. $F = 353.5$ lb;
$= \dfrac{250(\sin 125.0°)}{\sin 35.4°}$

101. $A = 333.8$;
$A = \dfrac{1}{2}(27.3)(35.2)(\sin 136.0°)$

103. $\theta = 2.4°$;
$\tan \theta = \dfrac{60^2}{86,400}$
$= 0.0417$

14.2 The Law of Sines

1. $\angle C = 72.6° = 180 - 70.8 - 36.6;$

   $b = 4.52, c = 7.23;$

   $\dfrac{\sin 70.8°}{7.16} = \dfrac{\sin 36.6°}{b} = \dfrac{\sin 72.6°}{c}$

3. $\angle C = 109.0°$

   $= 180 - 37.0 - 34.0;$

   $a = 1393.9, b = 1295.2;$

   $\dfrac{2190}{\sin 109.0°} = \dfrac{a}{\sin 37.0°}$

   $\phantom{\dfrac{2190}{\sin 109.0°}} = \dfrac{b}{\sin 34.0°}$

5. $\angle C = 9.6°;$

   $\dfrac{\sin 20.7°}{155} = \dfrac{\sin C}{72.8},$

   $\sin C = 0.1660;$

   $\angle A = 149.7° = 180 - 20.7 - 9.6;$

   $a = 221.2;$

   $\dfrac{155}{\sin 20.7°} = \dfrac{a}{\sin 149.7°}$

7. $\angle B = 8.5°;$

   $\dfrac{\sin 143.2°}{0.926} = \dfrac{\sin B}{0.228},$

   $\sin B = 0.1475;$

   $\angle C = 28.3° = 180 - 143.2 - 8.5;$

   $c = 0.733;$

   $\dfrac{0.926}{\sin 143.2°} = \dfrac{c}{\sin 28.3°}$

9. $\angle A = 99.4° = 180 - 58.4 - 22.2;$

   $b = 55.1, c = 24.4;$

   $\dfrac{63.8}{\sin 99.4°} = \dfrac{b}{\sin 58.4°}$

   $\phantom{\dfrac{63.8}{\sin 99.4°}} = \dfrac{c}{\sin 22.2°}$

11. $\angle A = 68.1° = 180 - 47.4 - 64.5;$

    $a = 552.1, c = 537.1;$

    $\dfrac{438}{\sin 47.4°} = \dfrac{a}{\sin 68.1°} = \dfrac{c}{\sin 64.5°}$

13. $\dfrac{\sin 48.1°}{22.2} = \dfrac{\sin A}{26.2}$

    $\sin A = 0.8784$

    $\angle A = 61.5°$ or $\angle A = 118.5°$

    $\angle C = 70.4°$ $\angle C = 13.4°$

    $\quad = 180 - 48.1 - 61.5;$ $= 180 - 48.1 - 118.5$

    $c = 28.1,$ $c = 6.9,$

    $\dfrac{22.2}{\sin 48.1°} = \dfrac{c}{\sin 70.4°}$ $\dfrac{22.2}{\sin 48.1°} = \dfrac{c}{\sin 13.4°}$

15. $\dfrac{\sin 31.4°}{576} = \dfrac{\sin C}{730}$

    $\sin C = 0.6603$

    $\angle C = 41.3°$ or $\angle C = 138.7°$

    $\angle A = 107.3°$ $\angle A = 9.9°$

    $\quad = 180 - 31.4 - 41.3;$ $= 180 - 31.4 - 138.7$

    $a = 1055.5,$ $a = 190.1,$

    $\dfrac{576}{\sin 31.4°} = \dfrac{a}{\sin 107.3°}$ $\dfrac{576}{\sin 31.4°} = \dfrac{a}{\sin 9.9°}$

17. $C = 42.4°,$

    $\dfrac{\sin 69.1°}{94.2} = \dfrac{\sin C}{68.0}$

    $\sin C = 0.6744;$

    $B = 68.5°$

    $\quad = 180 - 69.1 - 42.4;$

    $b = 93.8$

    $\dfrac{94.2}{\sin 69.1°} = \dfrac{b}{\sin 68.5°}$

19. not possible;

$$\frac{\sin 30.4°}{1.43} = \frac{\sin B}{4.21}$$

$$\sin B = 1.4898$$

21. $C = 90°;$

$$\frac{\sin 30.0°}{100} = \frac{\sin C}{200}$$

$$\sin C = 1;$$

$$B = 60° = 180 - 30.0 - 90;$$

$$b = 173.2,$$

$$b = \sqrt{c^2 - a^2}$$

$$= \sqrt{200^2 - 100^2} = \sqrt{30,000}$$

23. not possible;

$$\frac{\sin 85.2°}{2.14} = \frac{\sin C}{6.73}$$

$$\sin C = 3.1338$$

25. $x = 15.6$ in, $y = 27.2$ in;

$$180 - 54.8 = 125.2,$$

$$180 - 125.2 - 28.0 = 26.8,$$

$$\frac{x}{\sin 28.0°} = \frac{15.0}{\sin 26.8°}$$

$$= \frac{y}{\sin 125.2°}$$

27. $x = 20,978.4$ m;

$$180 - 47 = 133,$$

$$180 - 42 - 133 = 5,$$

$$\frac{x}{\sin 133°} = \frac{2500}{\sin 5°}$$

29. $x = 19.7$ km;

$$180 - 45 = 135,$$

$$\frac{\sin 135°}{90.0} = \frac{\sin P}{75.0}$$

$$\sin P = 0.5893$$

$$P = 36.1°$$

$$180 - 135 - 36.1 = 8.9$$

$$\frac{90}{\sin 135°} = \frac{x}{\sin 8.9°}$$

31. $x = 28,977.5$ km;

$$180 - 88.9 = 91.1,$$

$$180 - 87.6 - 91.1 = 1.3,$$

$$\frac{658}{\sin 1.3°} = \frac{x}{\sin 87.6°}$$

33. three

14.3 The Law of Cosines

1. $c = 27.1,$

$$c^2 = 22.3^2 + 16.4^2 - 2(22.3)(16.4)\cos 87.5°$$

$$c^2 = 798.2;$$

$$B = 37.1° \text{ by Law of Sines;}$$

$$A = 55.4° = 180 - 27.1 - 37.1$$

3. $b = 38,582.9,$

$$b^2 = 7720^2 + 42,000^2 - 2(7720)(42,000)\cos 58.9°$$

$$b^2 = 1,488,636,867;$$

$$A = 9.9°,$$

$$\cos A = \frac{42,000^2 + 38,582.9^2 - 7720^2}{2(42,000)(38,582.9)};$$

$$C = 111.2° = 180 - 58.9 - 9.9$$

5. $c = 1582.5$,

$c^2 = 1510^2 + 308^2 - 2(1510)(308)\cos 98.0°$

$c^2 = 2,504,417.3$;

$A = 70.9°$,

$\cos A = \dfrac{1582.5^2 + 308^2 - 1510^2}{2(1582.5)(308)}$;

$B = 11.1° = 180 - 98.0 - 70.9$

7. $C = 139.6°$,

$\cos C = \dfrac{934^2 + 770^2 - 1600^2}{2(934)(770)}$;

$A = 18.2°$,

$\cos A = \dfrac{934^2 + 1600^2 - 770^2}{2(934)(1600)}$;

$B = 22.2° = 180 - 18.2 - 139.6$

9. $C = 107.4°$,

$\cos C = \dfrac{51.2^2 + 38.4^2 - 72.6^2}{2(51.2)(38.4)}$;

$B = 30.3°$,

$\cos B = \dfrac{72.6^2 + 51.2^2 - 38.4^2}{2(72.6)(51.2)}$;

$A = 42.3° = 180 - 107.4 - 30.3$

11. $C = 93.4°$,

$\cos A = \dfrac{7.20^2 + 5.30^2 - 9.19^2}{2(7.20)(5.30)}$;

$B = 35.1°$,

$\cos B = \dfrac{9.19^2 + 7.20^2 - 5.30^2}{2(9.19)(7.20)}$;

$A = 51.5° = 180 - 93.4 - 35.1$

13. $c = 138.1$,

$c^2 = 238^2 + 312^2 - 2(238)(312)\cos 24.7°$

$c^2 = 19,063.6$;

$A = 46.1°$,

$\cos A = \dfrac{138.1^2 + 312^2 - 238^2}{2(138.1)(312)}$;

$B = 109.2° = 180 - 24.7 - 46.1$

15. $b = 4.20$,

$b^2 = 5.12^2 + 1.24^2 - 2(5.12)(1.24)\cos 37.3°$

$b^2 = 17.7$;

$A = 132.4°$,

$\cos A = \dfrac{1.24^2 + 4.20^2 - 5.12^2}{2(1.24)(4.20)}$;

$C = 10.3° = 180 - 37.3 - 132.5$

17. $C = 83.4°$,

$\cos C = \dfrac{831^2 + 637^2 - 987^2}{2(831)(637)}$;

$A = 39.9°$,

$\cos A = \dfrac{831^2 + 987^2 - 637^2}{2(831)(987)}$;

$B = 83.4° = 180 - 39.9 - 56.7$

19. $B = 102.3°$,

$\cos B = \dfrac{2.25^2 + 3.47^2 - 4.52^2}{2(2.25)(3.47)}$;

$C = 29.1°$,

$\cos C = \dfrac{3.47^2 + 4.52^2 - 2.25^2}{2(3.47)(4.52)}$;

$A = 48.6° = 180 - 29.1 - 102.3$

21. $C = 75.1°$,

$$\cos C = \frac{0.1287^2 + 0.1034^2 - 0.1429^2}{2(0.1287)(0.1034)};$$

$A = 44.4°$,

$$\cos A = \frac{0.1429^2 + 0.1287^2 - 0.1034^2}{2(0.1429)(0.1287)};$$

$B = 60.5° = 180 - 44.4 - 75.1$

23. $a = 24.25$,

$$a^2 = 34.26^2 + 10.21^2 - 2(34.26)(10.21)\cos 9.65°$$

$$a^2 = 588.6;$$

$B = 4.0°$,

$$\cos B = \frac{34.26^2 + 24.25^2 - 10.21^2}{2(34.26)(24.25)};$$

$C = 166.35° = 180 - 9.65 - 4.0$

25. $AC = b = 1287.9$;

$180 - 35 = 145$,

$$b^2 = 730^2 + 620^2 - 2(730)(620)\cos 145°$$

$$b^2 = 1,658,796$$

27. $a = 96.2$ cm;

$$a^2 = 56.0^2 + 65.0^2 - 2(56.0)(65.0)\cos 105.0°$$

$$a^2 = 9245.2;$$

29. $A = 19.8°$,

$$\cos A = \frac{17.85^2 + 11.55^2 - 8.00^2}{2(17.85)(11.55)}$$

31. $B = 90.0°$,

$$\cos B = \frac{4.58^2 + 7.93^2 - 9.16^2}{2(4.58)(7.93)};$$

$C = 30.0°$,

$$\cos C = \frac{7.93^2 + 9.16^2 - 4.58^2}{2(7.93)(9.16)};$$

$A = 60.0° = 180 - 30.0 - 90.0$

14.4 Introduction To Vectors

1. a. scalar--only magnitude is given
    b. vector--both magnitude and direction are given

3. a. vector--both magnitude and direction are given
    b. scalar--only magnitude is given

5. a. vector--both magnitude and direction are given
    b. scalar--only magnitude is given

7. a. scalar--only magnitude is given
    b. vector--both magnitude and direction are given

9.

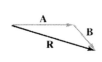

11.

13.

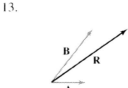

15.

17.

19.

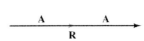

21.

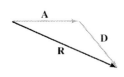

23.

25.

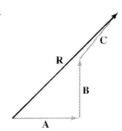

27.

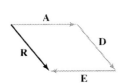

29.

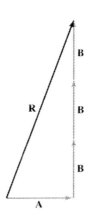

31.

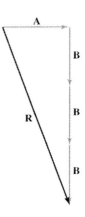

33.

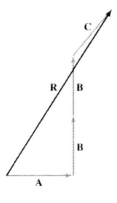

35.

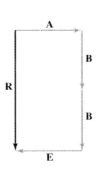

37.

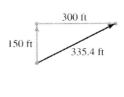

39.

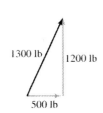

41.

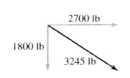

43.

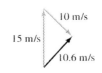

14.5 Vector Components

1.

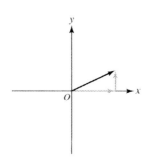

3.

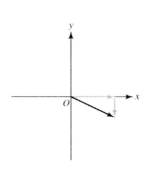

5. $x = 25.4$
$\quad = 28.5 \cos 27.0°;$
$y = 12.9$
$\quad = 28.5 \sin 27.0°$

7. $x = -8.6$
$\quad = 58.2 \cos 98.5°;$
$y = 57.6$
$\quad = 58.2 \sin 98.5°$

9. $x = 3437.3$
$\quad = 6560 \cos 301.6°;$
$y = -5587.3$
$\quad = 6560 \sin 301.6°$

11. $x = -700.5$
$\quad = 737 \cos 198.1°;$
$y = -229.0$
$\quad = 737 \sin 198.1°$

13. $x = -1343.7$
$\quad = 3482 \cos 247.3°;$
$y = -3212.3$
$\quad = 3482 \sin 247.3°$

15. $x = -0.13$
$\quad = 0.8734 \cos 98.6°;$
$y = 0.86$
$\quad = 0.8734 \sin 98.6°$

17. $x = 18.24$
$\quad = 25.8 \cos 45.0°;$
$y = 18.24$
$\quad = 25.8 \sin 45.0°$

19. $x = 3.45$
$\quad = 6.34 \cos 57.0°;$
$y = 5.32$
$\quad = 6.34 \sin 57.0°$

21. $x = -26.7$
$\quad = 219 \cos 263.0°;$
$y = -217.4$
$\quad = 219 \sin 263.0°$

23. $x = -968.7$
$\quad = 1370 \cos 225.0°;$
$y = -968.7$
$\quad = 1370 \sin 225.0°$

25. $x = 52.4$
$\quad = 87.2 \cos 306.9°;$
$y = -69.7$
$\quad = 87.2 \sin 306.9°$

27. $x = -0.032$
$\quad = 0.05436 \cos 233.5°;$
$y = -0.044$
$\quad = 0.05436 \sin 233.5°$

29. $x = 25,815.1$
$\quad = 25,872 \cos 356.20°;$
$y = -1714.6$
$\quad = 25,872 \sin 356.20°$

31. $x = -13.378$
$\quad = 34.666 \cos 112.70°;$
$y = 31.981$
$\quad = 34.666 \sin 112.70°$

33. $x = 33.3$
$\quad = 36.5 \cos 24.0°;$
$y = 14.8$
$\quad = 36.5 \sin 24.0°$

35. $x = 0.013$ m
$\quad = 0.024 \cos 58.3°;$
$y = 0.020$ m
$\quad = 0.024 \sin 58.3°$

37. $c = 3.72 \text{ A}$,

$$c = \sqrt{I_c{}^2 + I_R{}^2}$$

$$= \sqrt{2.35^2 + 2.89^2}\,;$$

$\theta = 39.1°$,

$$\tan\theta = \frac{I_c}{I_R} = \frac{2.35}{2.89}$$

14.6 Vector Addition

1. $r = 5.4$

$$= \sqrt{5^2 + 2^2}\,;$$

$\theta = 21.8°$,

$$\tan\theta = \frac{2.00}{5.00}$$

3. $r = 1455.7$

$$= \sqrt{746^2 + 1250^2}\,;$$

$\theta = 59.2°$,

$$\tan\theta = \frac{1250}{746}$$

5. $r = 38.3$

$$= \sqrt{34.7^2 + 16.3^2}\,;$$

$\theta = 25.2°$,

$$\tan\theta = \frac{16.3}{34.7}$$

7. $r = 25.5$

$$= \sqrt{22.46^2 + 12.05^2}\,;$$

$\theta = 61.8°$,

$$\tan\theta = \frac{22.46}{12.05}$$

9. $r = 10.74, \theta = 23.7°; x = 9.83, y = 4.32;$

$A_x = 7.63\cos 0° = 7.63$

$Bx = 4.85\cos 63.0° = 2.20$

$A_y = 7.63\sin 0° = 0$

$By = 4.85\sin 63.0° = 4.32$

$$r = \sqrt{9.83^2 + 4.32^2}\,,$$

$$\tan\theta = \frac{4.32}{9.83}$$

11. $r = 275.6, \theta = 55.3°; x = 157.1, y = 226.5;$

$A_x = 127\cos 21.0° = 118.6$

$Bx = 185\cos 78.0° = 38.5$

$A_y = 127\sin 21.0° = 45.5$

$By = 185\sin 78.0° = 181.0$

$$r = \sqrt{157.1^2 + 226.5^2}\,,$$

$$\tan\theta = \frac{226.5}{157.1}$$

13. $r = 115.0, \theta = 102.0°; x = -23.9, y = 112.5;$

$A_x = 95.4\cos 78.4° = 19.2$

$Bx = 47.1\cos 156.2° = -43.1$

$A_y = 95.4\sin 78.4° = 93.5$

$By = 47.1\sin 156.2° = 19.0$

$$r = \sqrt{(-23.9)^2 + 112.5^2}\,,$$

$$\tan\phi = \frac{112.5}{23.9}, \phi = 78.0, \theta = 180 - 78.0$$

15. $r = 120.6, \theta = 272.1°; x = 4.4, y = -120.5;$

   $A_x = 86.3 \cos 207.0° = -76.9$

   $Bx = 115 \cos(-45.0)° = 81.3$

   $A_y = 86.3 \sin 207.0° = -39.2$

   $By = 115 \sin(-45.0)° = -81.3$

   $r = \sqrt{(-120.5)^2 + 4.4^2},$

   $\tan \phi = \dfrac{120.5}{4.4}, \phi = 87.9, \theta = 360 - 87.9$

17. $r = 10, \theta = 36.9°; x = 8.0, y = 6.0;$

   $A_x = 8.0 \cos 0° = 8.0$

   $Bx = 6.0 \cos 90° = 0$

   $A_y = 8.0 \sin 0° = 0$

   $By = 6.0 \sin 90° = 6.0$

   $r = \sqrt{8.0^2 + 6.0^2},$

   $\tan \theta = \dfrac{6.0}{8.0}$

19. $r = 61.1, \theta = 116.7°; x = -27.5, y = 54.6;$

   $A_x = 47.2 \cos 90.0° = 0$

   $Bx = 28.5 \cos 165.0° = -27.5$

   $A_y = 47.2 \sin 90.0° = 47.2$

   $By = 28.5 \sin 165.0° = 7.4$

   $r = \sqrt{(-27.5)^2 + 54.6^2},$

   $\tan \phi = \dfrac{54.6}{27.5}, \phi = 63.3, \theta = 180 - 63.3$

21. $r = 1558.5, \theta = 201.6°; x = -1448.8, y = -574.3;$

   $A_x = 566 \cos 155.0° = -513.0$

   $Bx = 1240 \cos 221.0° = -935.8$

   $A_y = 566 \sin 155.0° = 239.2$

   $By = 1240 \sin 221.0° = -813.5$

   $r = \sqrt{(-1448.8)^2 + (-574.3)^2},$

   $\tan \phi = \dfrac{574.3}{1448.8}, \phi = 21.6, \theta = 180 + 21.6$

23. $r = 7675.1, \theta = 175.1°; x = -7646.7, y = 659.5;$

   $A_x = 1324.5 \cos 193.70° = -1286.8$

   $Bx = 6433.9 \cos 171.30° = -6359.9$

   $A_y = 1324.5 \sin 193.70° = -313.7$

   $By = 6433.9 \sin 171.30° = 973.2$

   $r = \sqrt{(-7646.7)^2 + 659.5^2},$

   $\tan \phi = \dfrac{659.5}{7646.7}, \phi = 4.9, \theta = 180 - 4.9$

25. $r = 13.0, \theta = 292.6°; x = 5.0, y = -12.0;$

   $A_x = 5.0 \cos 0° = 5.0$

   $Bx = 12.0 \cos(-90)° = 0$

   $A_y = 5.0 \sin 0° = 0$

   $By = 12.0 \sin(-90)° = -12.0$

   $r = \sqrt{(-12.0)^2 + 5.0^2},$

   $\tan \phi = \dfrac{12.0}{5.0}, \phi = 67.4, \theta = 360 - 67.4$

27. $r = 29.0, \theta = 327.4°; x = 24.4, y = -15.6;$

   $A_x = 16.4 \cos(-108.5)° = -5.20$

   $Bx = 29.6 \cos 0° = 29.6$

   $A_y = 16.4 \sin(-108.5)° = -15.6$

   $By = 29.6 \sin 0° = 0$

   $r = \sqrt{(-15.6)^2 + 24.4^2},$

   $\tan \phi = \dfrac{15.6}{24.4}, \phi = 32.6, \theta = 360 - 32.6$

29. $r = 4352.1, \theta = 321°; x = 3383.6, y = -2737.1;$

   $A_x = 2560 \cos(-32.6)° = 2156.7$

   $Bx = 1830 \cos(-47.9)° = 1226.9$

   $A_y = 2560 \sin(-32.6)° = -1379.3$

   $By = 1830 \sin(-47.9)° = -1357.8$

   $r = \sqrt{(-2737.1)^2 + 3383.6^2},$

   $\tan \phi = \dfrac{2737.1}{3383.6}, \phi = 39.0, \theta = 360 - 39.0$

31. $r = 0.321, \theta = 193.5°; x = -0.312, y = -0.075;$
$A_x = 0.859 \cos 156.2° = -0.786$
$Bx = 0.635 \cos(-41.7)° = 0.474$
$A_y = 0.859 \sin 156.2° = 0.347$
$By = 0.635 \sin(-41.7)° = -0.422$
$r = \sqrt{(-0.312)^2 + (-0.075)^2},$
$\tan \phi = \dfrac{0.075}{0.312}, \phi = 13.5, \theta = 180 + 13.5$

33. $\theta = 53.3°$ from the
158-lb force,
$r = 264.4$ lb;
$r = \sqrt{212^2 + 158^2}$
$\tan \theta = \dfrac{212}{158}$

35. $r = 46.0$ mi,
$\theta = 34.4°$ S of E;
$r = \sqrt{38.0^2 + 26.0^2}$
$\tan \theta = \dfrac{26}{38}$

37. $r = 3990.9, \theta = 5.0°; x = 3975.7, y = 347.9;$
$A_x = 1500 \cos 0° = 1500$
$Bx = 2500 \cos 8° = 2475.7$
$A_y = 1500 \sin 0° = 0$
$By = 2500 \sin 8° = 347.9$
$r = \sqrt{3975.7^2 + 347.9^2},$
$\tan \theta = \dfrac{347.9}{3975.7}$

39. $r = 53.9, \theta = 68.2°; x = 20.0, y = -50.0;$
$A_x = 50.0 \cos 270° = 0$
$Bx = 20.0 \cos 0° = 20.0$
$A_y = 50.0 \sin 270° = -50.0$
$By = 20.0 \sin 0° = 0$
$r = \sqrt{(-50.0)^2 + 20.0^2},$
$\tan \theta = \dfrac{50.0}{20.0}$

41. $r = 130.0, \theta = 6.8°; x = 129.1, y = 15.5;$
$A_x = 120.0 \cos 0° = 120$
$Bx = 18.0 \cos 59.7° = 9.1$
$A_y = 120.0 \sin 0° = 0$
$By = 18.0 \sin 59.7° = 15.5$
$r = \sqrt{129.1^2 + 15.5^2},$
$\tan \theta = \dfrac{15.5}{129.1}$

43. $A_x = 110 \cos 30° = 95.3$
$Bx = 75 \cos 20° = 70.5$
$\Rightarrow$ The object will move
to the right.
$A_y = 110 \sin 30° = 55$
$By = 75 \sin 20° = 25.7$
$\Rightarrow$ The object will not
move upward.

Review Exercises Chapter 14

1. $r = 5 = \sqrt{4^2 + 3^2}$

$\sin \theta = \dfrac{3}{5}$ $\csc \theta = \dfrac{5}{3}$

$\cos \theta = \dfrac{4}{5}$ $\sec \theta = \dfrac{5}{4}$

$\tan \theta = \dfrac{3}{4}$ $\cot \theta = \dfrac{4}{3}$

3. $r = \sqrt{53} = \sqrt{7^2 + (-2)^2}$

$\sin \theta = \dfrac{-2}{\sqrt{53}} = -0.2742$ $\csc \theta = \dfrac{\sqrt{53}}{-2} = -3.6401$

$\cos \theta = \dfrac{7}{\sqrt{53}} = 0.9615$ $\sec \theta = \dfrac{\sqrt{53}}{7} = 1.0400$

$\tan \theta = \dfrac{-2}{7} = -0.2857$ $\cot \theta = \dfrac{7}{-2} = -3.5$

5. II

7. IV

9. $-\cos 48°, \tan 14°$

11. $-\sin 71°, \sec 15°$

13. -0.4540

15. -0.3057

17. -1.0824

19. -0.5764

21. -2.5517

23. 4.2303

25. 0.6820

27. 1.0032

29. $\theta = 37.0°, 217.0°$

31. $\theta = 114.9°, 245.1°$

33. $\theta = 27.4°, 152.6°$

35. $\theta = 189.4°, 350.6°$

37. $\theta = 155.0°, 335.0°$

39. $\theta = 56.3°, 123.7°$

41. $C = 71.7°$
$= 180 - 61.3 - 47.0;$
$b = 119.9, c = 129.8$
$$\frac{100}{\sin 47.0°} = \frac{b}{\sin 61.3°} = \frac{c}{\sin 71.7°}$$

43. $A = 21.2°$
$= 180 - 148.8 - 10.0;$
$b = 128.5, c = 43.1$
$$\frac{89.7}{\sin 21.2°} = \frac{b}{\sin 148.8°} = \frac{c}{\sin 10.0°}$$

45. $c = 5.60,$
$c^2 = 3.20^2 + 4.50^2 - 2(3.20)(4.50)\cos 91.7°$
$c^2 = 31.3;$
$A = 34.8°,$
$$\frac{\sin 91.7°}{5.60} = \frac{\sin A}{3.20}$$
$\sin A = 0.5712;$
$B = 53.5°$
$= 180 - 91.7 - 34.8$

47. $b = 289.0,$
$b^2 = 283^2 + 278^2 - 2(283)(278)\cos 62.0°$
$b^2 = 83,502.6;$
$C = 58.1°,$
$$\frac{\sin 62.0°}{289.0} = \frac{\sin C}{278}$$
$\sin C = 0.8493;$
$A = 59.9°$
$= 180 - 62.0 - 58.1$

49. $\dfrac{\sin 54.3°}{16.3} = \dfrac{\sin C}{18.2}$
$\sin C = 0.9067$

$C = 65.1°$ or $C = 114.9°$
$A = 180 - 54.3 - 65.1$ $A = 180 - 54.3 - 114.9$
$A = 60.6°$ $A = 10.8°$
$\dfrac{\sin 54.3°}{16.3} = \dfrac{\sin 60.6°}{a}$ $\dfrac{16.3}{\sin 54.3°} = \dfrac{a}{\sin 10.8°}$
$a = 17.5$ $a = 3.8$

51. $\dfrac{\sin 51.0°}{65.5} = \dfrac{\sin B}{78.4}$

$\sin B = 0.9302$

$B = 68.5°$ or $B = 111.5°$

$C = 180 - 51.0 - 68.5$ $C = 180 - 51.0 - 111.5$

$C = 60.5°$ $C = 17.5°$

$\dfrac{65.5}{\sin 51.0°} = \dfrac{c}{\sin 60.5°}$ $\dfrac{65.5}{\sin 51.0°} = \dfrac{c}{\sin 17.5°}$

$c = 73.4$ $c = 25.3$

53. $A = 83.3°,$

$\cos A = \dfrac{289^2 + 184^2 - 324^2}{2(289)(184)};$

$B = 34.3°,$

$\dfrac{\sin 83.3°}{324} = \dfrac{\sin B}{184};$

$C = 62.4°$

$= 180 - 83.3 - 34.3$

55. $A = 133.1°,$

$\cos A = \dfrac{1.28^2 + 4.16^2 - 5.12^2}{2(1.28)(4.16)};$

$B = 10.5°,$

$\dfrac{\sin 133.1°}{5.12} = \dfrac{\sin B}{1.28};$

$C = 36.4°$

$= 180 - 133.1 - 10.5$

57. $a = 431.1,$

$a^2 = 243^2 + 295^2 - 2(243)(295)\cos 106.1°$

$a^2 = 185,832.6;$

$B = 32.8°,$

$\dfrac{\sin 106.1°}{431.1} = \dfrac{\sin B}{243};$

$C = 41.1°$

$= 180 - 106.1 - 32.8$

59. a. scalar--only magnitude is given

b. vector--both magnitude and direction are given

61. a. scalar--only magnitude is given

b. vector--both magnitude and direction are given

63.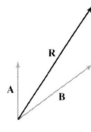

65.

67.

69.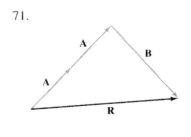

71.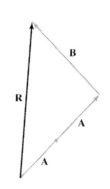

73.

75. $x = 6.77$
$= 7.30\cos 22.0°;$
$y = 2.73$
$= 7.30\sin 22.0°$

77. $x = -1375.2$
$= 1580\cos 150.5°;$
$y = 778.0$
$= 1580\sin 150.5°$

79. $A_x = 15.9$
$= 17.2\cos 22.2°;$
$A_y = 6.5$
$= 17.2\sin 22.2°$

81. $A_x = 8.90$
$= 16.48\cos(-57.3°);$
$A_y = -13.87$
$= 16.48\sin(-57.3°)$

83. $180 - 15 = 165,$
$x = -34.8$
$= 36\cos 165°;$
$y = 9.3$
$= 36\sin 165°$

85. $90 - 39 = 51,$
$x = 349.3$
$= 555\cos 51°;$
$y = 431.3$
$= 555\sin 51°$

87. $r = 38.7, \theta = 27.7°; x = 34.3, y = 18.0;$
$A_x = 12.0\cos 18.0° = 11.4$
$B_x = 27.0\cos 32.0° = 22.9$
$A_y = 12.0\sin 18.0° = 3.7$
$B_y = 27.0\sin 32.0° = 14.3$
$r = \sqrt{34.3^2 + 18.0^2},$
$\tan\theta = \dfrac{18.0}{34.3}$

89. $r = 5528.3, \theta = 81.7°; x = 798.6, y = 5470.3;$
$A_x = 2860\cos(-36.9°) = 2287.1$
$B_x = 7340\cos 101.7° = -1488.5$
$A_y = 2860\sin(-36.9°) = -1717.2$
$B_y = 7340\sin 101.7° = 7187.5$
$r = \sqrt{798.6^2 + 5470.3^2},$
$\tan\theta = \dfrac{5470.3}{798.6}$

91. $r = 27.66, \theta = 40.0°; x = 21.20, y = 17.76;$
$A_x = 17.76\cos 90.00° = 0$
$B_x = 21.20\cos 0.00° = 21.20$
$A_y = 17.76\sin 90.00° = 17.76$
$B_y = 21.20\sin 0.00° = 0$
$r = \sqrt{21.20^2 + 17.76^2},$
$\tan\theta = \dfrac{17.76}{21.20}$

93. $r = 1.100, \theta = 48.0°; x = 0.736, y = 0.818;$
$A_x = 0.992\cos 18.4° = 0.941$
$B_x = 0.545\cos 112.1° = -0.205$
$A_y = 0.992\sin 18.4° = 0.313$
$B_y = 0.545\sin 112.1° = 0.505$
$r = \sqrt{0.736^2 + 0.818^2},$
$\tan\theta = \dfrac{0.818}{0.736}$

95. $r = 47.0, \theta = 314.0°; x = 32.6, y = -33.8;$
$A_x = 36\cos 25° = 32.6$
$B_x = 49\cos(-90°) = 0$
$A_y = 36\sin 25° = 15.2$
$B_y = 49\sin(-90°) = -49$
$r = \sqrt{32.6^2 + (-33.8)^2},$
$\tan\phi = \dfrac{33.8}{32.6}, \phi = 46.0°, \theta = 360 - 46.0$

97. $r = 64.8, \theta = 68.7°; x = 23.6, y = 60.4;$
$A_x = 75.6\cos 29.4° = 65.9$
$B_x = 48.3\cos 151.2° = -42.3$
$A_y = 75.6\sin 29.4° = 37.1$
$B_y = 48.3\sin 151.2° = 23.3$
$r = \sqrt{23.6^2 + 60.4^2},$
$\tan\theta = \dfrac{60.4}{23.6}$

99. $A = 69.1°,$

$$\cos A = \frac{0.402^2 + 0.512^2 - 0.526^2}{2(0.402)(0.512)};$$

$B = 45.5°,$

$$\cos B = \frac{0.512^2 + 0.526^2 - 0.402^2}{2(0.512)(0.526)};$$

$C = 65.4°$

$= 180 - 69.1 - 45.5$

101. $v = -114.9$

$= 150 \cos 140.0°$

103. $x = 585.2$ m;

$B = 180 - 110 - 34.5$

$= 35.5°,$

$$\frac{600}{\sin 35.5°} = \frac{x}{\sin 34.5°}$$

105. $x = 2.2°;$

$$\sin x = \frac{3.10}{80.0}$$

107. $a = 281.0$ m;

$a^2 = 156^2 + 207^2 - 2(156)(207) \cos 100.5°$

$a^2 = 78{,}954.5$

109. $h = 2787.4$ ft;

$180 - 39.0 = 141.0,$

$180 - 38.2 - 141.0 = 0.8,$

$$\frac{100}{\sin 0.8°} = \frac{x}{\sin 38.2°}$$

$x = 4429.2,$

$$\sin 39.0° = \frac{h}{4429.2}$$

111. 113.4 lb;

$90 - 47 = 43,$

$155 \cos 43° = 113.4$

113. $x = 348.1$ km/h

$= 2250 \cos 81.1°$

$y = 2222.9$ km/h

$= 2250 \sin 81.1°$

115. $x = 231.9$ lb;

$= 305 \cos 40.5°$

117. $r = 1470.5$ lb, $\theta = 234.7°;$

$r = \sqrt{1200^2 + 850^2},$

$$\tan \phi = \frac{1200}{850}$$

$\phi = 54.7°$

119. $r = 88.3$ lb, $\theta = 81.1°$ from the

13.7-lb force;

$r = \sqrt{87.2^2 + 13.7^2},$

$$\tan \theta = \frac{87.2}{13.7}$$

121. $x = 32.4$ lb;

$= 36 \cos 26°$

123. $r = 356.4$ lb, $\theta = 23.1°$

from the 221-lb force; $x = 327.7$, $y = 140.0$,

$A_x = 221 \cos 0° = 221$

$B_x = 176 \cos 52.7° = 106.7$

$A_y = 221 \sin 0° = 0$

$B_y = 176 \sin 52.7° = 140.0$

$r = \sqrt{327.7^2 + 140.0^2}$, $\tan \theta = \dfrac{140.0}{327.7}$

Chapter 14 Test

1. $r = 10 = \sqrt{8^2 + 6^2}$

$\sin\theta = \dfrac{6}{10} = \dfrac{3}{5}$ $\csc\theta = \dfrac{5}{3}$

$\cos\theta = \dfrac{8}{10} = \dfrac{4}{5}$ $\sec\theta = \dfrac{5}{4}$

$\tan\theta = \dfrac{6}{8} = \dfrac{3}{4}$ $\cot\theta = \dfrac{4}{3}$

2. II

3. $C = 64° = 180 - 48 - 68;$

$b = 180.9, c = 175.4$

$\dfrac{145}{\sin 48°} = \dfrac{b}{\sin 68°} = \dfrac{c}{\sin 64°}$

4. $A = 67.0°,$

$\cos A = \dfrac{40.0^2 + 45.5^2 - 47.4^2}{2(40.0)(45.5)};$

$B = 51.0°,$

$\dfrac{\sin 67.0°}{47.4} = \dfrac{\sin B}{40.0};$

$C = 62.0°$

$= 180 - 67.0 - 51.0$

5. $x = 57.4$

$= 65\cos 28°$

$y = 30.5$

$= 65\sin 28°$

6. $r = 320.6, \theta = 113.4°; x = -127.2, y = 294.3,$

$A_x = 449\cos 74.2° = 122.3$

$B_x = 285\cos 208.9° = -249.5$

$A_y = 449\sin 74.2° = 432.0$

$B_y = 285\sin 208.9° = -137.7$

$r = \sqrt{127.2^2 + 294.3^2},$

$\tan\phi = \dfrac{294.3}{127.2}, \phi = 66.6°, \theta = 180 - 66.6$

7. $x = 24334$ km;

$180 - 89.2 = 90.8,$

$180 - 86.5 - 90.8 = 2.7,$

$\dfrac{1290}{\sin 2.7°} = \dfrac{x}{\sin 86.5°}$

8. $AC = 30.8$ m, $BC = 85.6$ m;

$180 - 74.3 - 85.4 = 20.3,$

$\dfrac{88.6}{\sin 85.4°} = \dfrac{AC}{\sin 20.3°} = \dfrac{BC}{\sin 74.3°}$

9. $r = 6.60$ lb, $\theta = 60.5°$ from the 3.25-lb force;

$r = \sqrt{5.75^2 + 3.25^2},$

$\tan\theta = \dfrac{5.75}{3.25}$

10. $r = 24.0$ m, $\theta = 36.5°; x = 19.3, y = 14.3;$

$A_x = 27.3\cos 45° = 19.3$

$B_x = 5\cos(-90)° = 0$

$A_y = 27.3\sin 45° = 19.3$

$B_y = 5\sin(-90)° = -5$

$r = \sqrt{19.3^2 + 14.3^2},$

$\tan\theta = \dfrac{14.3}{19.3}$

Chapter 15
Graphs of Trigonometric Functions

15.1 Radian Measure

1. $\dfrac{2\pi}{9}; \dfrac{40°}{1} \times \dfrac{\pi}{180}$

   $\dfrac{4\pi}{45}; \dfrac{16°}{1} \times \dfrac{\pi}{180}$

3. $\dfrac{11\pi}{36}; \dfrac{55°}{1} \times \dfrac{\pi}{180}$

   $\dfrac{11\pi}{6}; \dfrac{330°}{1} \times \dfrac{\pi}{180}$

5. $\dfrac{\pi}{6}; \dfrac{30°}{1} \times \dfrac{\pi}{180}$

   $\dfrac{3\pi}{4}; \dfrac{135°}{1} \times \dfrac{\pi}{180}$

7. $\dfrac{35\pi}{36}; \dfrac{175°}{1} \times \dfrac{\pi}{180}$

   $\dfrac{7\pi}{6}; \dfrac{210°}{1} \times \dfrac{\pi}{180}$

9. $120°; \dfrac{2\pi}{3} \times \dfrac{180}{\pi}$

   $36°; \dfrac{\pi}{5} \times \dfrac{180}{\pi}$

11. $15°; \dfrac{\pi}{12} \times \dfrac{180}{\pi}$

    $225°; \dfrac{5\pi}{4} \times \dfrac{180}{\pi}$

13. $168°; \dfrac{14\pi}{15} \times \dfrac{180}{\pi}$

    $140°; \dfrac{7\pi}{9} \times \dfrac{180}{\pi}$

15. $35°; \dfrac{7\pi}{36} \times \dfrac{180}{\pi}$

    $135°; \dfrac{3\pi}{4} \times \dfrac{180}{\pi}$

17. $0.802; 46.0° \times \dfrac{\pi}{180}$

19. $3.32; 190° \times \dfrac{\pi}{180}$

21. $4.86; 278.6° \times \dfrac{\pi}{180}$

23. $3.18; 182.4° \times \dfrac{\pi}{180}$

25. $45.8°; 0.80 \times \dfrac{180}{\pi}$

27. $143.2°; 2.5 \times \dfrac{180}{\pi}$

29. $186.2°; 3.25 \times \dfrac{180}{\pi}$

31. $710.5°; 12.4 \times \dfrac{180}{\pi}$

33. $\dfrac{\pi}{3} = 60°,$

    $\cos 60° = 0.5$

35. $\dfrac{2\pi}{3} = 120°,$

    $\tan 120° = -1.7321$

37. $1.05 = 60.2°,$

    $\sin 60.2° = 0.8674$

39. $0.875 = 50.1°,$

    $\tan 50.1° = 1.1960$

41. $\dfrac{5\pi}{18} = 50°,$

    $\sec 50° = 1.5557$

43. $3.00 = 171.9°,$

    $\csc 171.9° = 7.0972$

45. 0.8573

47. -1.1448

49. -9.6569

51. 3.7958

53. $\theta = 1.0472, 5.2360$

55. $\theta = 1.9116, 5.0532$

57. $\theta = 0.4500, 2.6916$

59. $\theta = 4.4704, 4.9543$

61. $v = 161.7$ V;
    $v = 170\sin[377(0.0050)]$

63. $H = 33.4$ m;
    $H = 37.4\cos 0.465$

65. $4.5 \times 360 = 1620°$,
    $4.5 \times 2\pi = 9\pi$ rad

67. $\dfrac{2.55(180)}{\pi} = 146.10°$

69. $\dfrac{\pi(180)}{13\pi} = 13.8°$

71. $\dfrac{75\pi}{180} = \dfrac{5\pi}{12}$

73. $\dfrac{10\pi}{180} = \dfrac{\pi}{18}$

75. -0.3152

77. 0.0004210

15.2 Applications of Radian Measure

1. a. $s = 7.34$ cm;
       $= (3.65)(2.01)$
    b. $A = 13.39$ cm^2;
       $= \dfrac{1}{2}(2.01)(3.65)^2$

3. $\theta = 49.3° \times \dfrac{\pi}{180} = 0.86$
    a. $s = 354.32$ mm;
       $= (412)(0.86)$
    b. $A = 72{,}989.92$ mm^2;
       $= \dfrac{1}{2}(0.86)(412)^2$

5. a. $s = 0.62$ ft;
       $= (2.37)\left(\dfrac{\pi}{12}\right)$
    b. $A = 0.74$ ft^2;
       $= \dfrac{1}{2}\left(\dfrac{\pi}{12}\right)(2.37)^2$

7. $\theta = 235° \times \dfrac{\pi}{180} = \dfrac{47\pi}{36}$
    a. $s = 26.86$ in;
       $= (6.55)\left(\dfrac{47\pi}{36}\right)$
    b. $A = 87.98$ in^2;
       $= \dfrac{1}{2}\left(\dfrac{47\pi}{36}\right)(6.55)^2$

9. $v = 70.63$ cm/s;
    $= (5.65)12.5)$

11. $v = 6924.57$ ft/min;
    $= (1.28)(861)2\pi$

13. $\theta = \dfrac{60.0° \times \pi}{180} = \dfrac{\pi}{3}$
    $s = 37.70$ cm
    $= (36.0)\left(\dfrac{\pi}{3}\right)$

15. $A = 678.58$ cm^2
    $= \dfrac{1}{2}\left(\dfrac{\pi}{3}\right)(36.0)^2$

17. $\theta = 1.62$;
    $s = r\theta$
    $118 = 73.0\theta$

19. 87.84 ft^2;

$$\theta = \frac{60°\pi}{180} = \frac{\pi}{3},$$

$$A_L = \frac{1}{2}\left(\frac{\pi}{3}\right)(18.0)^2 = 169.65 \text{ ft}^2,$$

$$A_S = \frac{1}{2}\left(\frac{\pi}{3}\right)(12.5)^2 = 81.81 \text{ ft}^2,$$

$$A_L - A_S = 87.84$$

21. 0.1134 m;

$$\theta = \frac{6.2\pi}{180} = 0.108,$$

$$s = (1.05)(0.108)$$

23. 26.39 ft;

$$\theta = \frac{84.0\pi}{180} = \frac{7\pi}{15},$$

$$s = (180)\left(\frac{7\pi}{15}\right)$$

25. $\omega = 58.67$ r/s;

$$r = \frac{33.0}{2} = 16.5 \text{ in},$$

$$\frac{55 \text{ mi}}{\text{hr}} \times \frac{5280 \text{ ft}}{1 \text{ mi}} \times \frac{12 \text{ in}}{1 \text{ ft}} \times \frac{1 \text{ hr}}{3600 \text{ sec}} = \frac{968 \text{ in}}{\text{sec}},$$

$$v = \omega r$$

$$968 = \omega(16.5)$$

$$\omega = 58.67$$

27. 152,681.4 cm/min;

$$r = \frac{54.0}{2} = 27.0;$$

$$v = (900)(27.0)2\pi$$

29. 134.04 ft^2;

$$\theta = \frac{240\pi}{180} = \frac{4\pi}{3},$$

$$r = \frac{16}{2} = 8.0,$$

$$A = \frac{1}{2}\left(\frac{4\pi}{3}\right)(8.0)^2$$

31. $1'' = \frac{1}{60}' = \frac{1}{3600}° \times \frac{\pi}{180}$

$$= 0.000004848,$$

$$\sin(0.000004848) = 0.000004848$$

$$\tan(0.000004848) = 0.000004848$$

33. 80,355,656.9 mi;

$$\theta = 0.000054° \times \frac{\pi}{180} = 0.000000942,$$

$$s = (5.88 \times 10^{12} \times 14.5)(0.000000942)$$

15.3 The Graphs of $y=a \sin x$ and $y=a \cos x$

1.

x	$-\pi$	$\dfrac{-3\pi}{4}$	$\dfrac{-\pi}{2}$	$\dfrac{-\pi}{4}$	0	$\dfrac{\pi}{4}$	$\dfrac{\pi}{2}$	$\dfrac{3\pi}{4}$	π	$\dfrac{5\pi}{4}$	$\dfrac{3\pi}{2}$	$\dfrac{7\pi}{4}$	2π	$\dfrac{9\pi}{4}$	$\dfrac{5\pi}{2}$	$\dfrac{11\pi}{4}$	3π
$\sin x$	0	$\dfrac{-1}{\sqrt{2}}$	-1	$\dfrac{-1}{\sqrt{2}}$	0	$\dfrac{1}{\sqrt{2}}$	1	$\dfrac{1}{\sqrt{2}}$	0	$\dfrac{-1}{\sqrt{2}}$	-1	$\dfrac{-1}{\sqrt{2}}$	0	$\dfrac{1}{\sqrt{2}}$	1	$\dfrac{1}{\sqrt{2}}$	0

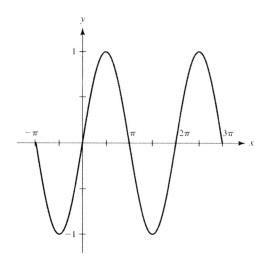

3.

x	$-\pi$	$\dfrac{-3\pi}{4}$	$\dfrac{-\pi}{2}$	$\dfrac{-\pi}{4}$	0	$\dfrac{\pi}{4}$	$\dfrac{\pi}{2}$	$\dfrac{3\pi}{4}$	π	$\dfrac{5\pi}{4}$	$\dfrac{3\pi}{2}$	$\dfrac{7\pi}{4}$	2π	$\dfrac{9\pi}{4}$	$\dfrac{5\pi}{2}$	$\dfrac{11\pi}{4}$	3π
$5\sin x$	0	$\dfrac{-5}{\sqrt{2}}$	-5	$\dfrac{-5}{\sqrt{2}}$	0	$\dfrac{5}{\sqrt{2}}$	5	$\dfrac{5}{\sqrt{2}}$	0	$\dfrac{-5}{\sqrt{2}}$	-5	$\dfrac{-5}{\sqrt{2}}$	0	$\dfrac{5}{\sqrt{2}}$	5	$\dfrac{5}{\sqrt{2}}$	0

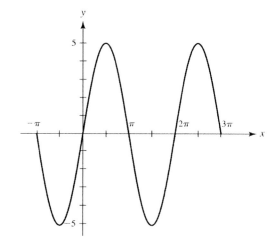

5. $y = 3 \sin x$

7. $y = -6 \sin x$

9.

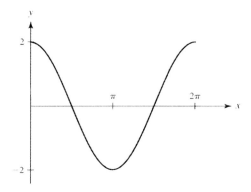

11.

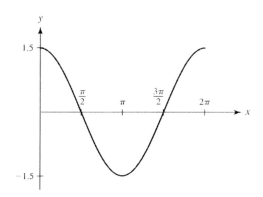

13.

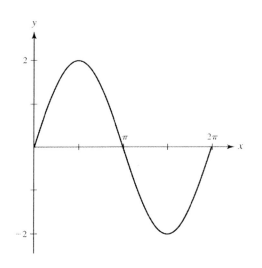

15.

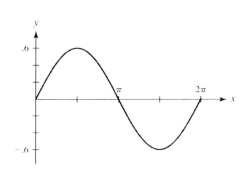

17.

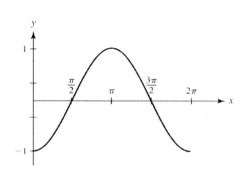

19.

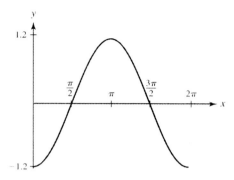

21.

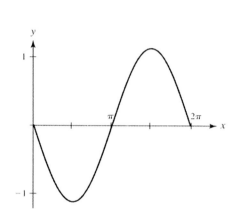

23.

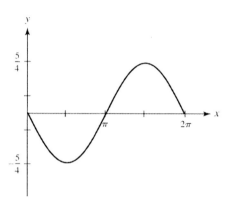

25.

x	0	1	2	3	4	5	6	7
$\cos x$	1	0.5403	-0.4161	-0.9900	-0.6536	0.2837	0.9602	0.7539

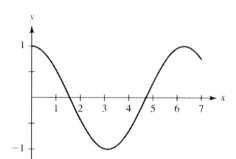

27. x	0	1	2	3	4	5	6	7
$\sin x$	0	0.8415	0.9093	0.1411	-0.7568	-0.9589	-0.2794	0.6570

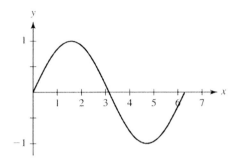

29. x	0	1	2	3	4	5	6	7
$\dfrac{2}{3}\sin x$	0	0.5610	0.6062	0.0941	-0.5045	-0.6393	-0.1863	0.4380

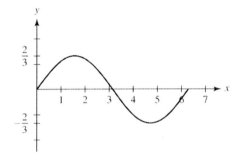

31. x	0	1	2	3	4	5	6	7
$-2\cos x$	-2	-1.0806	0.8323	1.9800	1.3073	-0.5673	-1.9203	-1.5078

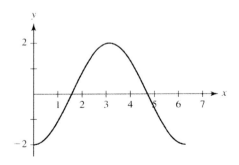

33.

x	0	1	2	3	4	5	6	7
$-2.6\cos x$	-2.6	-1.4048	1.0820	2.5740	1.6995	-0.7375	-2.4964	-1.9601

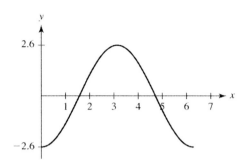

35.

x	0	1	2	3	4	5	6	7
$-3.5\sin x$	0	-2.9451	-3.1825	-0.4939	-2.6488	-3.3562	0.9780	-2.2995

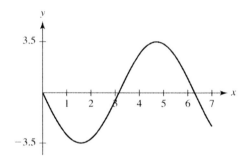

15.4 Graphs of $y = a \sin bx$ and $y = a \cos bx$

1. $\dfrac{2\pi}{3}$

3. $\dfrac{2\pi}{4} = \dfrac{\pi}{2}$

5. $\dfrac{2\pi}{3}$

7. $\dfrac{2\pi}{6} = \dfrac{\pi}{3}$

9. $\dfrac{2\pi}{5}$

11. $\dfrac{2\pi}{2\pi} = 1$

13. $\dfrac{2\pi}{1/2} = 4\pi$

15. $\dfrac{2\pi}{2/3} = 3\pi$

17. $\dfrac{2\pi}{3\pi/2} = \dfrac{4}{3}$

19. $\dfrac{2\pi}{\pi^2} = \dfrac{2}{\pi}$

21.

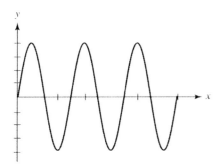

23.

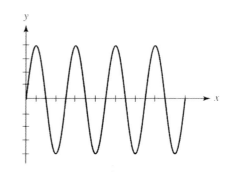

25.

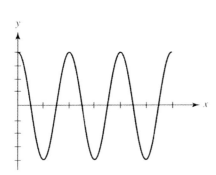

27.

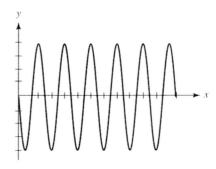

29.

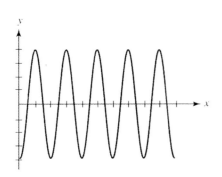

31.

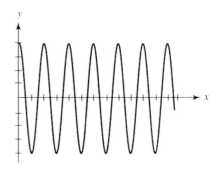

33.

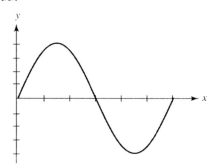

35.

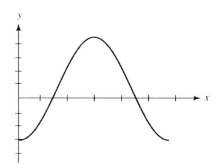

37.

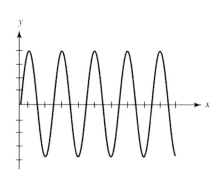

39.

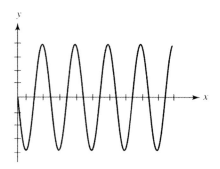

41. $y = \sin 2x$

43. $y = \sin \pi x$

45. $y = \sin 2\pi x$

47. $y = \sin 5x$

49. $y = 2\sin \pi x$

51. $y = 5\sin \dfrac{1}{2}\pi x$

53.

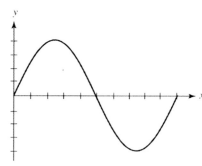

55.

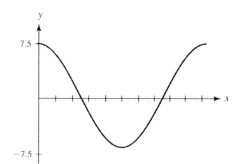

57.

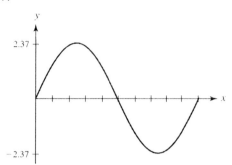

59.

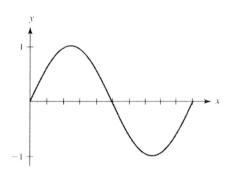

15.5 Graphs of $y = a \sin(bx + c)$ and $y = a \cos(bx + c)$

1. $A = 1, \text{per} = 2\pi,$

   $\text{displ} = \dfrac{-\pi}{3}$

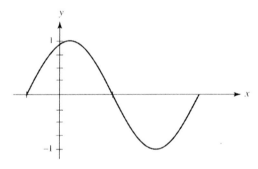

3. $A = 1, \text{per} = 2\pi,$

   $\text{displ} = \dfrac{\pi}{3}$

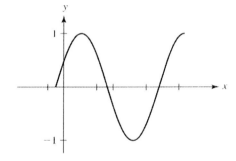

5. $A = 3, \text{per} = \pi,$

   $\text{displ} = \dfrac{-\pi}{8}$

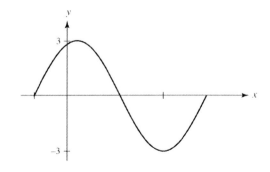

7. $A = 4, \text{per} = \dfrac{2\pi}{3},$

   $\text{displ} = \dfrac{-\pi}{3}$

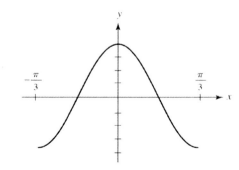

9. $A = 2, \text{per} = 6\pi,$

    $\text{displ} = \dfrac{-3\pi}{2}$

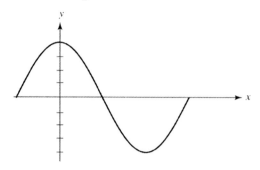

11. $A = 6, \text{per} = 3\pi,$

    $\text{displ} = \dfrac{-\pi}{2}$

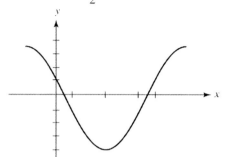

13. $A = 10, \text{per} = 2,$

    $\text{displ} = \dfrac{1}{\pi}$

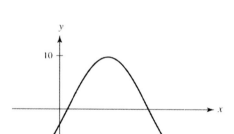

15. $A = \dfrac{3}{2}, \text{per} = \dfrac{2}{3},$

    $\text{displ} = \dfrac{1}{9\pi}$

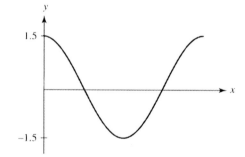

17. $A = 1.2, \text{per} = 2,$

    $\text{displ} = \dfrac{1}{6}$

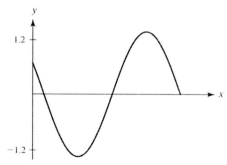

19. $A = 6, \text{per} = 2,$

    $\text{displ} = \dfrac{1}{2\pi}$

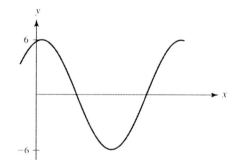

21. $A = \dfrac{3}{4}$, per $= \dfrac{2}{\pi}$,

  displ $= \dfrac{-1}{\pi}$

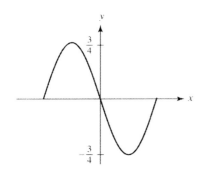

23. $A = \dfrac{5}{2}$, per $= 2$,

  displ $= \dfrac{\pi}{3}$

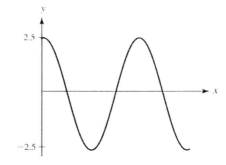

25.

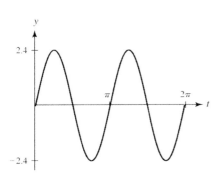

27.

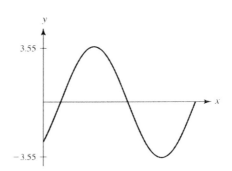

15.6 The Sine Function as a Function of Time

1.

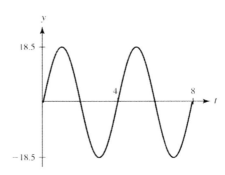

3.

5.

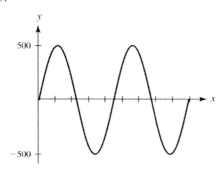

7.

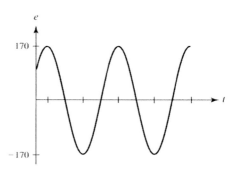

9.

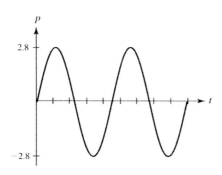

11. max current = 25 A,

$$\text{period} = \frac{2\pi}{635},$$

$$f = \frac{635}{2\pi}$$

13.

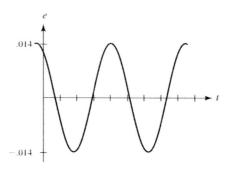

15.

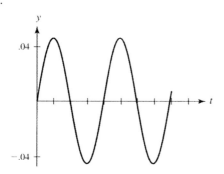

Review Exercises Chapter 15

1. $\dfrac{63.0\pi}{180} = \dfrac{7\pi}{20}$;

   $\dfrac{137\pi}{180}$

3. $\dfrac{46.0\pi}{180} = \dfrac{23\pi}{90}$;

   $\dfrac{10\pi}{180} = \dfrac{\pi}{18}$

5. $\dfrac{\pi}{9}\left(\dfrac{180}{\pi}\right) = 20°$;

   $\dfrac{5\pi}{6}\left(\dfrac{180}{\pi}\right) = 150°$

7. $\dfrac{7\pi}{18}\left(\dfrac{180}{\pi}\right) = 70°$;

   $\dfrac{9\pi}{20}\left(\dfrac{180}{\pi}\right) = 81°$

9. $0.625\left(\dfrac{180}{\pi}\right) = 35.8°$

11. $3.45\left(\dfrac{180}{\pi}\right) = 197.7°$

13. $75°\left(\dfrac{\pi}{180}\right) = 1.31$

15. $340.0°\left(\dfrac{\pi}{180}\right) = 5.93$

17. $152.5°\left(\dfrac{\pi}{180}\right) = 2.66$

19. $9.3°\left(\dfrac{\pi}{180}\right) = 0.16$

21. $\theta = 0.6894, 2.4522$

23. $\theta = 1.9984, 4.2848$

25. $\theta = 35.7°\left(\dfrac{\pi}{180}\right) = 0.6231$

   a. $s = 7.73$ cm

   $= (12.4)(0.6231)$

   b. $A = 47.90$ cm^2

   $= \dfrac{1}{2}(0.6231)(12.4)^2$

27. a. $s = 2.27$ in

   $= (7.24)\left(\dfrac{\pi}{10}\right)$

   b. $A = 8.23$ in^2

   $= \dfrac{1}{2}\left(\dfrac{\pi}{10}\right)(7.24)^2$

29. $v = 4.60$ m/min

   $= (12.6)(0.365)$

31. $v = 353.43$ m/min

   $= (45.0)(1.25)2\pi$

33.

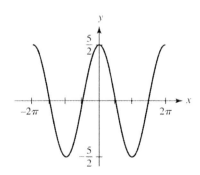

35.

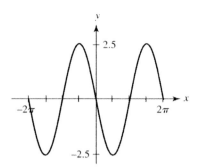

37.

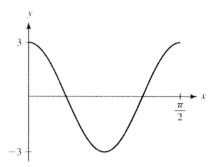

39.

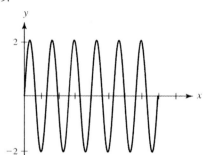

41.

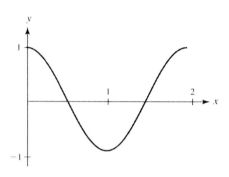

43.

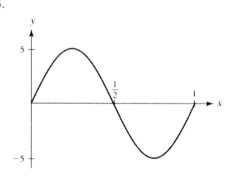

45.

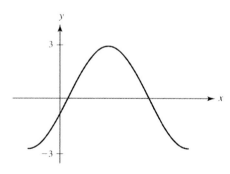

47.

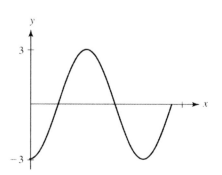

49.

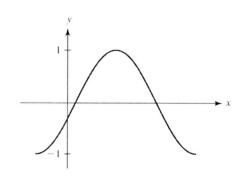

51.

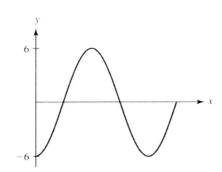

53. $y = 8\sin \pi x$

55. $y = -3\cos 2\pi x$

57. $d = 8.78$ cm

$= 12.0\cos[5(0.150)]$

59. $\theta = 2.04$ rad;

$A = \pi r^2(0.325)$ and

$A = \dfrac{1}{2}\theta r^2$, so

$\dfrac{1}{2}\theta r^2 = 0.325\pi r^2$

$\dfrac{1}{2}\theta = 0.325\pi$

$\theta = 2(0.325)\pi$

61. $A = 36.3$ cm^2

$= \dfrac{1}{2}\left(\dfrac{8.50}{2}\right)^2\left(\dfrac{230.5\pi}{180}\right)$

63. $\omega = 21.9$;

$v = (35.0)(3.60)$

$= 126$,

$126 = \omega(5.75)$

$\omega = \dfrac{126}{5.75}$

65. $\omega = 5688.9$ r/min;

$r = \dfrac{45.0 \text{ cm}}{2} \div 100 = 0.225$ m,

$1280 = \omega(0.225)$

$\omega = \dfrac{1280}{0.225}$

67.

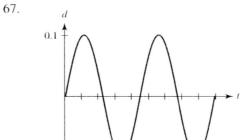

69. $v = 21.5$ V

$= 120\sin[36(0.0050)]$

71.

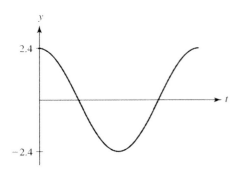

Chapter 15 Test

1. $150°\left(\dfrac{\pi}{180}\right) = \dfrac{5\pi}{6}$

2. $\dfrac{2\pi}{5}\left(\dfrac{180}{\pi}\right) = 72°$

3. $3.572\left(\dfrac{180}{\pi}\right) = 204.7°$

4. $s = 3.14$

   $= (1.50)\left(\dfrac{4\pi}{6}\right)$

5. $A = 32,383.4 \text{ cm}^2$

   $= \dfrac{1}{2}(0.803)(284)^2$

6. $v = 38,704.4 \text{ ft/min}$

   $= (2200)(2.80)(2\pi)$

7.

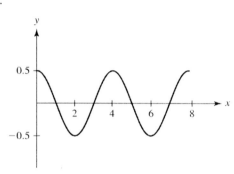

8.

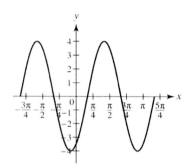

9.

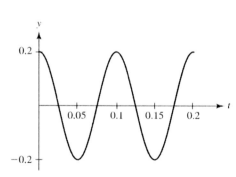

Chapter 16
Complex Numbers

16.1 Introduction to Complex Numbers

1. $4j$

3. $-3j$

5. $0.5j$

7. $3j\sqrt{3}$

9. $-4j\sqrt{3}$

11. $0.02j$

13. $5+j$

15. $6-2j$

17. $-4+2j\sqrt{2}$

19. $14-3j\sqrt{7}$

21. $-9; (3j)^2 = 9j^2$

23. $-6; \left(j\sqrt{6}\right)^2 = 6j^2$

25. $\sqrt{9} = 3$

27. $-2\sqrt{3}; \left(j\sqrt{3}\right)(2j) = 2j^2\sqrt{3}$

29. $-6; \left(j\sqrt{3}\right)\left(2j\sqrt{3}\right) = 6j^2$

31. $-\sqrt{33}; \left(j\sqrt{3}\right)\left(j\sqrt{11}\right) = j^2\sqrt{33}$

33. $-1; j^4 j^2 = j^2 = -1$

35. $j; j^8 j = j$

37. $j; j^{64} j = j$

39. $-1; j^{400} j^2 = j^2 = -1$

41. $-1; (-j)^{22} = j^{22} = j^2 = -1$

43. $-j; -j^{81} = -j^{80} j = -j$

45. $3+8j$

47. -4

49. $-9j$

51. $-2+7j$

16.2 Basic Operations With Complex Numbers

1. $5-7j$

3. $4+5j$

5. $1+3j$

7. $-12j$

9. $2-6j$

11. $-3+20j$

13. $-30; 30 j^2$

15. $5;$
$$4 + 2j - 2j - j^2$$
$$4 - (-1)$$

17. $48 - 9j;$
$$6 + 12j - 21j - 42j^2$$
$$6 - 9j + 42$$

19. $36j;$
$$-36j^3$$
$$(-36)(-j)$$

21. $6 + 4j;$
$$4j - 6j^2$$

23. $36 + 3j;$
$$6 - 12j + 15j - 30j^2$$
$$6 + 3j + 30$$

25. $\dfrac{2j}{3-4j} \times \dfrac{3+4j}{3+4j} = \dfrac{6j + 8j^2}{9 - 16j^2}$
$$= \dfrac{-8 + 6j}{25}$$

27. $\dfrac{8-3j}{2}; = 4 - \dfrac{3}{2}j$

$\dfrac{3+8j}{2j} \times \dfrac{-2j}{-2j} = \dfrac{-6j - 16j^2}{-4j^2}$
$$= \dfrac{16 - 6j}{4}$$

29. $\dfrac{53 - 27j}{61};$

$\dfrac{7+3j}{5+6j} \times \dfrac{5-6j}{5-6j} = \dfrac{35 - 42j + 15j - 18j^2}{25 - 36j^2}$
$$= \dfrac{35 - 27j + 18}{25 + 36}$$

31. $\dfrac{-5-j}{2};$

$\dfrac{-4+7j}{1-3j} \times \dfrac{1+3j}{1+3j} = \dfrac{-4 - 12j + 7j + 21j^2}{1 - 9j^2}$
$$= \dfrac{-25 - 5j}{10}$$

33. $\dfrac{3+3j}{4};$

$\dfrac{3j}{2+2j} \times \dfrac{2-2j}{2-2j} = \dfrac{6j - 6j^2}{4 - 4j^2}$
$$= \dfrac{6 + 6j}{8}$$

35. $\dfrac{-3+7j}{6};$

$\dfrac{2+5j}{3-3j} \times \dfrac{3+3j}{3+3j} = \dfrac{6 + 6j + 15j + 15j^2}{9 - 9j^2}$
$$= \dfrac{-9 + 21j}{18}$$

37. $-2j;$
$$1 - j - j + j^2$$
$$1 - 2j - 1$$

39. $2 - 3j;$
$$(1+j)(1+j) + (2 - 5j)$$
$$(1 + j + j + j^2) + (2 - 5j)$$
$$2j + 2 - 5j$$

41. $-30 - 40j;$
$$5j(1+3j)^2$$
$$5j(1 + 3j + 3j + 9j^2)$$
$$5j(-8 + 6j)$$
$$-40j + 30j^2$$

43. $3 + 4j$;

$$\frac{(2+j)(3-j)}{1-j} = \frac{7+j}{1-j},$$

$$\frac{7+j}{1-j} \times \frac{1+j}{1+j} = \frac{7+7j+j+j^2}{1-j^2}$$

$$= \frac{6+8j}{2}$$

45. $10 + 6j$;

$$6j - (1+j)(1+2j)(1+3j)$$

$$6j - (-1+3j)(1+3j)$$

$$6j - (-10)$$

47. $-48j$;

$$(4-3j)^2 - (4+3j)^2$$

$$(16-24j+9j^2) - (16+24j+9j^2)$$

$$16-24j+9j^2 - 16-24j-9j^2$$

49. 100;

$$(6+8j)(6-8j)$$

$$36 - 64j^2$$

$$36 + 64$$

51. 53;

$$(-2+7j)(-2-7j)$$

$$4 - 49j^2$$

$$4 + 49$$

53. $z = 14.7 + j(10.2 - 12.3)$

$$= 14.7 - 2.1j \text{ ohms}$$

55. $V = 3.1 + 5.8j$ volts;

$$V = ZI$$

$$= (-0.20 + 1.3j)(4.0 - 3.0j)$$

$$= -0.8 + 0.6j + 5.2j - 3.9j^2$$

16.3 Graphical Representations of Complex Numbers

1.

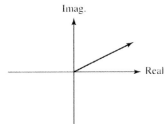

3.

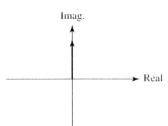

5.

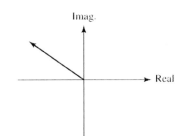

7.

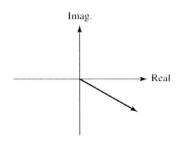

9. $5 + 4j$

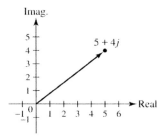

11. $2 + 6j$

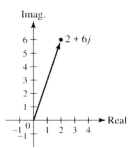

13. $6 + j$

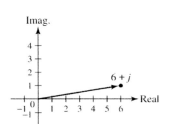

15. $7 + j$

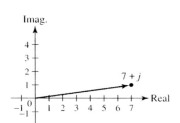

17. $4 - j$

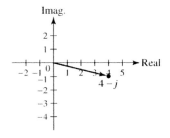

19. $9 - 15j$

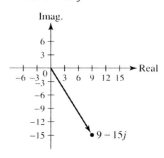

21. 4

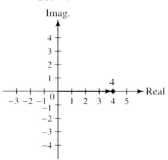

23. $-13 + 2j$

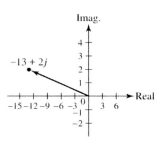

25. $-15 - 11j$

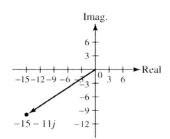

27. $12 - 4j$

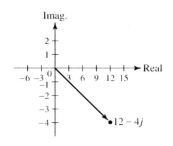

29.

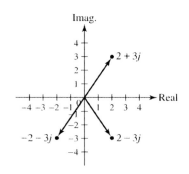

31.

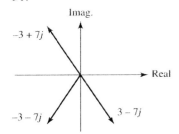

16.4 Polar Form of Complex Numbers

1. $5(\cos 36.9° + j \sin 36.9°)$;

$r = \sqrt{4^2 + 3^2}$

$= 5,$

$\tan \theta = \dfrac{3}{4}$

$\theta = 36.9°$

3. $5(\cos 323.1° + j \sin 323.1°)$;

$r = \sqrt{(-3)^2 + 4^2}$

$= 5,$

$\tan \phi = \dfrac{3}{4}$

$\phi = 36.9°$

$\theta = 360 - 36.9$

5. $4.50(\cos 302.9° + j \sin 302.9°)$;

$r = \sqrt{(-3.78)^2 + 2.45^2}$

$= 4.50,$

$\tan \phi = \dfrac{3.78}{2.45}$

$\phi = 57.1°$

$\theta = 360 - 57.1$

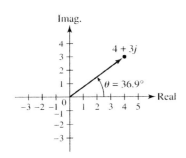

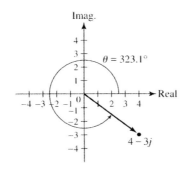

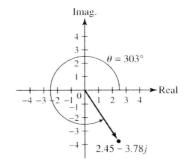

7. $6.52(\cos 55.8° + j \sin 55.8°);$

$r = \sqrt{3.66^2 + 5.39^2}$

$= 6.52,$

$\tan \theta = \dfrac{5.39}{3.66}$

$\theta = 55.8°$

9. $2(\cos 225° + j \sin 225°);$

$r = \sqrt{\left(-\sqrt{2}\right)^2 + \left(-\sqrt{2}\right)^2}$

$= 2,$

$\tan \phi = \dfrac{\sqrt{2}}{\sqrt{2}}$

$\phi = 45°$

$\theta = 180 + 45$

11. $2(\cos 120° + j \sin 120°);$

$r = \sqrt{\left(-1\right)^2 + \sqrt{3}^2}$

$= 2,$

$\tan \phi = \dfrac{\sqrt{3}}{1}$

$\phi = 60°$

$\theta = 180 - 60$

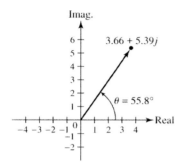

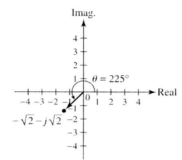

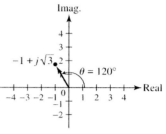

13. $10(\cos 180° + j \sin 180°);$

$r = \sqrt{\left(-10\right)^2 + 0^2}$

$= 10,$

$\tan \theta = \dfrac{0}{-10}$

$\theta = 180°$

15. $6(\cos 90° + j \sin 90°);$

$r = \sqrt{0^2 + 6^2}$

$= 6,$

$\tan \theta = \dfrac{6}{0}$

$\theta = 90°$

17. $6.15(\cos 43.0° + j \sin 43.0°)$

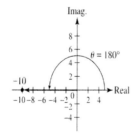

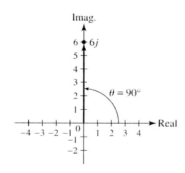

19. $10.3(\cos 335.2° + j \sin 335.2°)$

21. $0.348(\cos 76.4° + j \sin 76.4°)$

23. $56.0(\cos 212.5° + j \sin 212.5°)$

25. $1.03 + 2.82 j$

27. $1.17 - 0.94 j$

29. $-10 j$

31. -25

33. $40.96 + 28.68 j$

35. $6.2 - 10.74 j$

37. $10.06 + 7.38 j$

39. $-1.258 - 1.797 j$

41. $56.8 + 7.2 j$

43. $r = 81.2$

$\quad = \sqrt{64.8^2 + (-49.0)^2}$;

$\quad \theta = 322.9°,$

$\quad \tan \phi = \dfrac{49.0}{64.8}$

$\quad \phi = 37.1$

$\quad \theta = 360 - 37.1$

Review Exercises Chapter 16

1. $8j$

3. $-20j$

5. $3j\sqrt{6}$

7. $-2j\sqrt{14}$

9. $-6 + 10j$

11. $3 - 4j\sqrt{3}$

13. $-25;$

$\quad 5j \times 5j = 25j^2$

15. $-\sqrt{14};$

$\quad j\sqrt{7} \times j\sqrt{2} = j^2\sqrt{14}$

17. $-1;$

$\quad j^{12}j^2 = j^2$

19. $1;$

$\quad (-j)^{40} = j^{40}$

21. $14 + 4j$

23. $-1 - 7j$

25. $15j;$

$\quad 6 + 3j + 12j + 6j^2$

$\quad 6 + 15j - 6$

27. $-11 + 27j;$

$\quad 9 + 15j + 12j + 20j^2$

$\quad 9 + 27j - 20$

29. $\quad\quad\quad \dfrac{-2 - 16j}{5};$

$\quad \dfrac{2 - 10j}{3 + j} \times \dfrac{3 - j}{3 - j} = \dfrac{6 - 2j - 30j + 10j^2}{9 - j^2}$

$\quad\quad\quad\quad\quad\quad = \dfrac{-4 - 32j}{10}$

31. $\quad\quad\quad \dfrac{-46 - 2j}{53};$

$\quad \dfrac{6 + 2j}{-7 - 2j} \times \dfrac{-7 + 2j}{-7 + 2j} = \dfrac{-42 + 12j - 14j + 4j^2}{49 - 4j^2}$

$\quad\quad\quad\quad\quad\quad = \dfrac{-46 - 2j}{53}$

33. $(6 + 8j) + (6 - 8j) = 12;$

$\quad (6 + 8j)(6 - 8j) = 36 - 64j^2$

$\quad\quad\quad\quad\quad\quad = 100$

35. $(7 - 3j) + (7 + 3j) = 14;$

$\quad (7 - 3j)(7 + 3j) = 49 - 9j^2$

$\quad\quad\quad\quad\quad\quad = 58$

37. $3 + 3j$

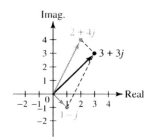

39. $-9 - 7j$

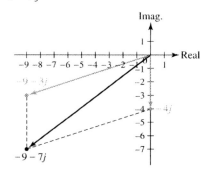

41. $13(\cos 67.4° + j\sin 67.4°)$

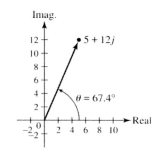

43. $21.8(\cos 214.3° + j\sin 214.3°)$

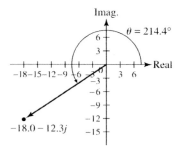

45. $5.1 + 14.3j$

47. $14.1 - 9.0j$

49. yes;

$(1 + 2j)^2 - 2(1 + 2j) + 5$

$1 + 4j + 4j^2 - 2 - 4j + 5$

$1 + 4j - 4 - 2 - 4j + 5$

$0;$

yes;

$(1 - 2j)^2 - 2(1 - 2j) + 5$

$1 - 4j + 4j^2 - 2 + 4j + 5$

$1 - 4j - 4 - 2 + 4j + 5$

0

51. $V = 9 + 42j$ volts;

$V = IZ$

$\quad = (5 - 4j)(-3 + 6j)$

$\quad = -15 + 30j + 12j - 24j^2$

Chapter 16 Test

1. $3+j$

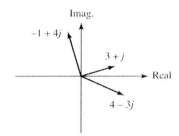

2. $3+4j$

3. $-22-14j$;

$$-10+6j-20j+12j^2$$
$$-10-14j-12$$

4. $\dfrac{-14-5j}{13}$;

$$\dfrac{-4+j}{3-2j}\times\dfrac{3+2j}{3+2j}=\dfrac{-12-8j+3j+2j^2}{9-4j^2}$$

5. $\sqrt{53}(\cos 285.9^\circ+j\sin 285.9^\circ)$;

$$r=\sqrt{2^2+(-7)^2},$$
$$\tan\phi=\frac{7}{2}$$
$$\phi=74.1$$
$$\theta=360-74.1$$

6. $-1.414-1.414j$

7. not a solution;

$$(5-2j)^2-2(5-2j)+4$$
$$25-20j+4j^2-10+4j+4$$
$$15-16j\neq 0$$

Chapter 17
Introduction to Data Analysis

17.1 Creating Pie Charts and Bar Graphs

1.

Deaths from Accidents

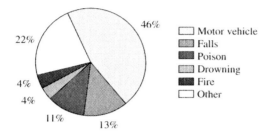

3.

Rainfall by Season

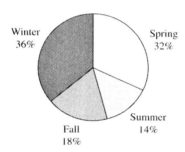

5.

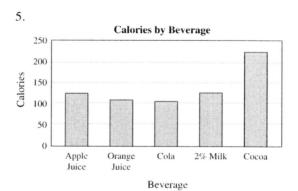

7.

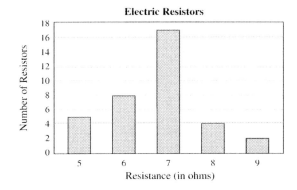

17.2 Frequency Tables and Histograms

1.

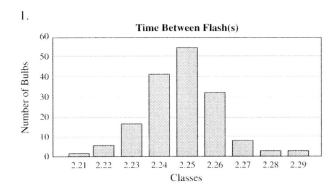

3.

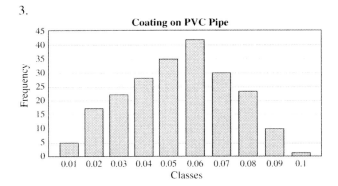

5.

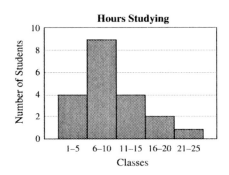

7.

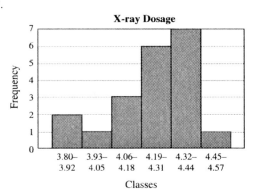

9.

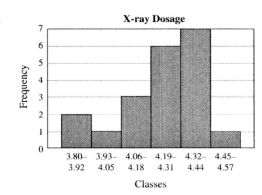

11.

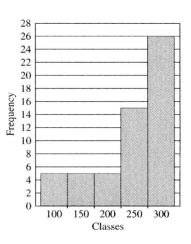

13.

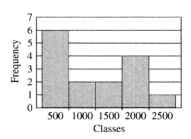

17.3 Measures of Central Tendency

1. mean $= \dfrac{\sum x}{n} = \dfrac{313}{15} = 21$

median $= 21$

mode $= 21$

3. mean $= \dfrac{\sum x}{n} = \dfrac{275.8}{18} = 15.3$

median $= \dfrac{14 + 14.6}{2} = 14.3$

mode $= 14, 9, 16$

5. mean $= \dfrac{\sum x}{n} = \dfrac{2404}{14} = 172$

mode $= 144$

7. mean $= \dfrac{\sum x}{n} = \dfrac{40313}{15} = 2688$

median $= 2749$

9. $63 - 26 = 37$

11. mean $= \dfrac{\sum x}{n} = \dfrac{14461}{15} = 964$

13. mean $= \dfrac{\sum x}{n} = \dfrac{3.70}{18} = 0.21$

   median $= 0.195$

   mode $= 0.18$

15. mean $= \dfrac{\sum x}{n} = \dfrac{11.49}{215} = 0.05$ mm

17.4 Measures of Spread and Variation

1. Range $= 25 - 18 = 7$

   Standard Deviation

   <u>Step 1</u> Find the mean
   Mean $= 313/15 = 20.9$

   <u>Step 2</u> Subtract the mean from each score $\left(x - \overline{x} \right)$

   $19 - 20.9 = -1.9$
   $21 - 20.9 = 0.1$
   $22 - 20.9 = 1.1$
   $25 - 20.9 = 4.1$
   $22 - 20.9 = 1.1$
   $20 - 20.9 = -0.9$
   $18 - 20.9 = -2.9$
   $21 - 20.9 = 0.1$
   $20 - 20.9 = -0.9$
   $19 - 20.9 = -1.9$
   $22 - 20.9 = 1.1$
   $21 - 20.9 = 0.1$
   $19 - 20.9 = -1.9$
   $23 - 20.9 = 2.1$
   $21 - 20.9 = 0.1$

   <u>Step 3</u> Square each value in Step 2 $\left(x - \overline{x} \right)^2$

   $3.61, 0.01, 1.21, 16.81, 1.21, 0.81, 8.41, 0.01, 0.81, 3.61, 1.21, 0.01, 3.61, 4.41, 0.01$

   <u>Step 4</u> Sum Step 3
   45.75

   <u>Step 5</u> Divide Step 4 by $n - 1$
   $\dfrac{45.75}{14} = 3.26786$

   <u>Step 6</u> Take the square root
   $\sqrt{3.26786} = 1.81$

3. Range = $4.1 - 3.3 = 0.8$
   Standard Deviation

   Step 1 Find the mean
   Mean = $60.2/16 = 3.8$

   Step 2 Subtract the mean from each score $\left(x - \overline{x}\right)$
   $3.7 - 3.8 = -0.1$
   $3.6 - 3.8 = -0.2$
   $3.6 - 3.8 = -0.2$
   $4 - 3.8 = 0.2$
   continue with rest of scores

   Step 3 Square each value in Step 2 $\left(x - \overline{x}\right)^2$
   $0.01, 0.04, 0.4, 0.4, 0.4, \ldots$

   Step 4 Sum Step 3
   0.76

   Step 5 Divide Step 4 by $n - 1$
   $\dfrac{0.76}{15} = 0.0507$

   Step 6 Take the square root
   $\sqrt{0.0507} = 0.225$

5. standard deviation is 0.18

7. standard deviation is 0.05

9. range is 2406; standard deviation is 825

11. range is 38.9; standard deviation is 11.4

13. Golfer 1 mean 81.1; range 13; standard deviation 3.2
    Golfer 2 mean 81.2; range 17; standard deviation 5.9
    Golfer 1 is more consistent

17.5 Probability

1. $\dfrac{435}{1500} = 0.29$

3. $\dfrac{435 + 345}{1500} = 0.52$

5. $\dfrac{231}{3300} = 0.07$

7. $\dfrac{3}{8} = 0.375$

9. $\dfrac{13}{52} = 0.25$

11. $\dfrac{300}{6000} = 0.05$

13. $\dfrac{3000 - 55}{3000} = 0.98$

15. $\dfrac{15 + 32}{120} = 0.39$

Review Exercises

1.

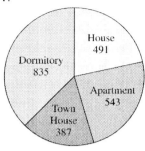

Dormitory 835
House 491
Apartment 543
Town House 387

3.

Classes	Frequency
1093–1095	3
1096–1098	4
1099–1101	3
1102–1104	4
1105–1107	2

5. mean $= \dfrac{17600}{16} = 1100$

7. mode $= 1098$

9. 4.066

11.

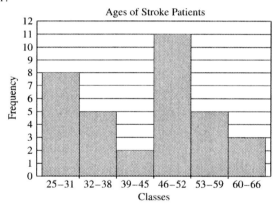

Ages of Stroke Patients

13. $\dfrac{46+48}{2} = 47$

15. $63 - 26 = 37$

17.

Classes	Frequency
11,000–17,999	7
18,000–24,999	17
25,000–31,999	16
32,000–38,999	8
39,000–45,999	2

19. 25,708

21. 20,400

23. 7337

25. $\dfrac{152}{580} = 0.26$

27. $\dfrac{300 - 55 - 98 - 60 - 10}{300} = 0.26$

Chapter Test

1.

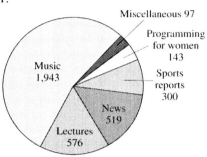

Miscellaneous 97

Programming for women 143

Music 1,943

Sports reports 300

News 519

Lectures 576

2.

Classes	Frequency
45–49	8
50–54	13
55–59	12
60–64	7
65–69	3

3.

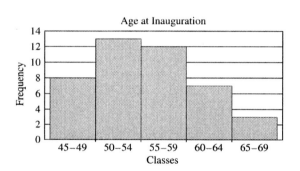

4. 54

5. 55

6. 51, 54

7. 27

8. 6.2

9. $P(54) = \dfrac{5}{43} = 0.12$

10. $P(60\text{'s}) = \dfrac{10}{43} = 0.23$

Appendix

A.1 Addition and Subtraction of Whole Numbers

1. 152

3. 1234

5. 2420

7. 21,410

9. 25,295

11. 139,639

13. 5602

15. 581

17. 949

19. 9798

21. 30,579

23. 95,682

A.2 Multiplication and Division of Whole Numbers

1. 10,534

3. 1,314,976

5. 1,048,576

7. 236,894,403

9. 244

11. 324

13. 2048

15. 1981 R 104

17. $646 \bullet 74 = 17 \bullet 2812$
$47804 = 47804$

19. $14670 \bullet 217 = 326 \bullet 9765$
$3183390 = 3183390$

21. $15 \bullet 3 + 15 \bullet 7 = 15 \bullet 10$
$45 + 105 = 150$
$150 = 150$

23. $18212 + 54008 = 628 \bullet 115$
$72220 = 72220$

25. 20 mi x 8 mi = 160 mi^2

27. 17 in x 14 in = 238 in^2

29. 296 cm x 35 cm = 10,360 cm^2

31. 18 in x 24 in = 432 in^2

33. 595 mi/h x 17 h = 10,115 mi

35. 8 yd x 5 yd = 4o yd^2; $32/yd^2 x 40 yd^2 = $1280

37. 43560 ft^2/132 ft = 330 ft

39. 234 mi/13 gal = 18 mi/gal

A.3 Fractions

1. $\frac{5}{9}$

3. $\frac{1}{7}$

5. $\frac{7}{13}; \frac{3}{16}$

7. $\frac{9}{8}; \frac{1}{12}$

9. 1, 6

11. 32, 1

13. $1\frac{2}{3}$

15. $4\frac{12}{13}$

17. $3\frac{53}{75}$

19. $53\frac{4}{25}$

21. $\frac{17}{5}$

23. $\frac{79}{8}$

25. $\frac{223}{13}$

27. $\frac{423}{4}$

29. $\frac{17}{24}$

31. $\frac{10}{23}$

33. $8\frac{1}{2}$ in

35. $\frac{37}{10}$ in

A.4 Equivalent Fractions

1. $\frac{3 \bullet 2}{7 \bullet 2} = \frac{6}{14}$

3. $\frac{16 \div 4}{20 \div 4} = \frac{4}{5}$

5. $\frac{4 \bullet 6}{13 \bullet 6} = \frac{24}{78}$

7. $\frac{60 \div 12}{156 \div 12} = \frac{5}{13}$

9. $\frac{13 \bullet 7}{25 \bullet 7} = \frac{91}{175}$

11. $\frac{1024 \div 32}{64 \div 32} = \frac{32}{2}$

13. $\frac{4}{8} = \frac{1}{2}$

15. $\frac{3}{2}$

17. $\frac{4}{5}$

19. $\frac{3}{5}$

21. $\frac{2}{5}$

23. $\frac{2}{5}$

25. $\frac{1 \bullet 4}{3 \bullet 4} = \frac{4}{12}$

27. $\frac{2 \bullet 5}{3 \bullet 5} = \frac{10}{15}$

29. $\frac{12 \div 4}{64 \div 4} = \frac{3}{16}$

31. $\frac{4}{6} = \frac{2 \bullet 2}{2 \bullet 3}$

33. $\frac{3 \bullet 2}{40 \bullet 2} = \frac{6}{80}$

35. $\frac{13 \bullet 4}{16 \bullet 4} = \frac{52}{64}$

37. 20 = 2 x 2 x 5

39. 16 = 2 x 2 x 2 x 2

41. 36 = 2 x 2 x 3 x 3

43. 48 = 2 x 2 x 2 x 2 x 3

45. $57 = 3 \times 19$

47. $105 = 3 \times 5 \times 7$

49. $\dfrac{30}{35} = \dfrac{2 \bullet 3 \bullet 5}{5 \bullet 7} = \dfrac{6}{7}$

51. $\dfrac{24}{30} = \dfrac{2 \bullet 2 \bullet 2 \bullet 3}{2 \bullet 3 \bullet 5} = \dfrac{4}{5}$

53. $\dfrac{56}{24} = \dfrac{2 \bullet 2 \bullet 2 \bullet 7}{2 \bullet 2 \bullet 2 \bullet 3} = \dfrac{7}{3}$

55. $\dfrac{52}{78} = \dfrac{2 \bullet 2 \bullet 13}{2 \bullet 3 \bullet 13} = \dfrac{2}{3}$

57. $\dfrac{17}{68} = \dfrac{17}{2 \bullet 2 \bullet 17} = \dfrac{1}{4}$

59. $\dfrac{63}{105} = \dfrac{3 \bullet 3 \bullet 7}{3 \bullet 5 \bullet 7} = \dfrac{3}{5}$

61. $\dfrac{12}{16}$ in $= \dfrac{3}{4}$ in

63. $\dfrac{14}{32} = \dfrac{7}{16}$

65. $\dfrac{18}{18 + 27} = \dfrac{18}{45} = \dfrac{2}{5}$ mL

67. $\dfrac{2000}{2400}; \dfrac{400}{2400}$

$\dfrac{5}{6}; \dfrac{1}{6}$

69. $\dfrac{25}{750} = \dfrac{1}{30}$

A.5 Addition and Subtraction of Fractions

1. 4

3. 24

5. 72

7. 300

9. $\dfrac{1}{5} + \dfrac{3}{5} = \dfrac{4}{5}$

11. $\dfrac{5}{7} - \dfrac{3}{7} = \dfrac{2}{7}$

13. $\dfrac{2}{4} + \dfrac{1}{4} = \dfrac{3}{4}$

15. $\dfrac{23}{24} - \dfrac{20}{24} = \dfrac{3}{24} = \dfrac{1}{8}$

17. $\dfrac{4}{12} + \dfrac{9}{12} = \dfrac{13}{12}$

19. $\dfrac{5}{10} - \dfrac{4}{10} = \dfrac{1}{10}$

21. $\dfrac{15}{3} - \dfrac{8}{3} = \dfrac{7}{3}$

23. $\dfrac{22}{40} + \dfrac{25}{40} = \dfrac{47}{40}$

25. $\dfrac{65}{78} - \dfrac{9}{78} = \dfrac{56}{78} = \dfrac{29}{39}$

27. $3\dfrac{20}{24} - \dfrac{9}{24} = 3\dfrac{11}{24} = \dfrac{83}{24}$

29. $\dfrac{24}{42} + \dfrac{14}{42} - \dfrac{9}{42} = \dfrac{29}{42}$

31. $\dfrac{27}{72} + \dfrac{42}{72} + \dfrac{4}{72} = \dfrac{73}{72}$

33. $\dfrac{14}{42} + \dfrac{69}{42} - \dfrac{10}{42} = \dfrac{73}{42}$

35. $\dfrac{45}{60} + \dfrac{26}{60} - \dfrac{35}{60} + \dfrac{120}{60} = \dfrac{156}{60} = \dfrac{13}{5}$

37. $\dfrac{4}{16}+\dfrac{1}{16}+\dfrac{3}{16}=\dfrac{8}{16}=\dfrac{1}{2}$ in

39. $7\dfrac{15}{20}+\dfrac{4}{20}=7\dfrac{19}{20}$ gal

41. $\dfrac{6}{24}+\dfrac{26}{24}+\dfrac{9}{24}=\dfrac{41}{24}=1\dfrac{17}{24}$ in

43. $34\dfrac{1}{3}+17\dfrac{1}{3}=51\dfrac{2}{3}$

$51\dfrac{2}{3}-43\dfrac{1}{2}=51\dfrac{4}{6}-43\dfrac{3}{6}=8\dfrac{1}{6}$ yd^3

A.6 Multiplication and Division of Fractions

1. $\dfrac{2}{7}\bullet\dfrac{3}{11}=\dfrac{6}{77}$

3. $\dfrac{3}{1}\bullet\dfrac{2}{5}=\dfrac{6}{5}$

5. $\dfrac{7}{2\bullet2\bullet2}\bullet\dfrac{2\bullet3}{5}=\dfrac{21}{20}$

7. $\dfrac{5}{3\bullet3}\bullet\dfrac{1}{3}=\dfrac{5}{27}$

9. $\dfrac{2\bullet2\bullet2}{3\bullet3}\bullet\dfrac{5}{2\bullet2\bullet2\bullet2}=\dfrac{5}{18}$

11. $\dfrac{7}{3}\bullet\dfrac{3\bullet3}{2\bullet7}=\dfrac{3}{2}$

13. $\dfrac{2\bullet2\bullet2}{3\bullet5}\bullet\dfrac{5\bullet7}{2\bullet2\bullet3}=\dfrac{14}{9}$

15. $\dfrac{2\bullet2\bullet2}{17}\bullet\dfrac{1}{2\bullet2}=\dfrac{2}{17}$

17. $\dfrac{3}{5}\bullet\dfrac{3\bullet5}{7}\bullet\dfrac{2\bullet7}{3\bullet3}=\dfrac{6}{3}=2$

19. $\dfrac{3}{2\bullet2}\bullet\dfrac{2\bullet2\bullet7}{3\bullet3\bullet3}\bullet\dfrac{2\bullet3}{5\bullet7}=\dfrac{2}{15}$

21. $\dfrac{2\bullet3}{11}\bullet\dfrac{13}{2\bullet2\bullet3}\bullet\dfrac{11\bullet11}{2\bullet13}=\dfrac{11}{4}$

23. $\dfrac{9}{16}\left(\dfrac{2}{4}+\dfrac{1}{4}\right)$

$\dfrac{3\bullet3}{2\bullet2\bullet2\bullet2}\bullet\dfrac{3}{2\bullet2}=\dfrac{27}{64}$

25. $\dfrac{2\bullet3}{1}\bullet\dfrac{2\bullet2\bullet7}{23}=\dfrac{168}{23}$

27. $\dfrac{3}{6}+\dfrac{2}{6}=\dfrac{5}{6}$

$\dfrac{5}{6}\bullet\dfrac{3}{2}=\dfrac{5}{4}$

29. $\dfrac{7}{6}\div\dfrac{2}{12}=\dfrac{7}{2\bullet3}\bullet\dfrac{2\bullet2\bullet3}{2}=7$

31. $\dfrac{47}{63}\div\dfrac{39}{28}=\dfrac{47}{3\bullet3\bullet7}\bullet\dfrac{2\bullet2\bullet7}{3\bullet13}=\dfrac{188}{351}$

33. $\dfrac{19}{20}\div\dfrac{2}{30}=\dfrac{19}{2\bullet2\bullet5}\bullet\dfrac{2\bullet3\bullet5}{2}=\dfrac{57}{4}$

35. $\dfrac{4}{11}\div\dfrac{19}{15}=\dfrac{2\bullet2}{11}\bullet\dfrac{3\bullet5}{19}=\dfrac{60}{209}$

37. $\dfrac{1}{5};\dfrac{1}{13}$

39. $2;5$

41. $\dfrac{3}{16};\dfrac{2}{7}$

43. $\dfrac{16}{81};\dfrac{10}{71}$

45. $16 \bullet \dfrac{9}{4} = 3x$

    $36 = 3x$

    $x = 12$

47. $\dfrac{9}{4} \bullet \dfrac{2}{13} = \dfrac{9}{26}$ acres/h

49. $\dfrac{161}{5} \bullet \dfrac{2}{5} = \dfrac{322}{25} = 12\dfrac{22}{25}$ ft/s^2

51. $24 \text{ ft} \div \dfrac{3}{4} \text{ ft} = 24 \bullet \dfrac{4}{3} = 32$

A.7 Decimals

1. $4(10) + 7(1) + \dfrac{3}{10}$

3. $4(100) + 2(10) + 9(1) + \dfrac{4}{10} + \dfrac{8}{100} + \dfrac{6}{1000}$

5. 27.3

7. 57.54

9. 8.03

11. 17.4

13. 0.4

15. 0.21

17. 1.7

19. 0.499

21. $\dfrac{8}{10} = \dfrac{4}{5}$

23. $\dfrac{45}{100} = \dfrac{9}{20}$

25. $\dfrac{534}{100} = \dfrac{267}{50}$

27. $\dfrac{252}{10000} = \dfrac{63}{2500}$

29. 31.295

31. 2817.256

33. 8.763

35. 0.02628

37. 13.99952

39. 9.882

41. $5.6(22.32) = 124.992$

43. $6.75(6.66207) = 44.9689725$

45. 12.5

47. 295.4

49. 5.20

51. 0.46

53. $118.2 - 117.6 = 0.6$ V

55. $416.34 - 238.85 = \$177.49$

57. $318.62 - 82.74 = 235.88$ L

59. 16500 mi/h x 1.58 h = 26,070 mi

61. $1340/742 = 1.8$

63. $157/0.139 = 1129$ kw/h

A.8 Percents

1. 0.08

3. 2.36

5. 0.003

7. 0.056

9. 27%

11. 321%

13. 0.64%

15. 700%

17. $\dfrac{30}{100} = \dfrac{3}{10}$

19. $\dfrac{25}{1000} = \dfrac{1}{40}$

21. $\dfrac{120}{100} = \dfrac{6}{5}$

23. $\dfrac{57}{10000}$

25. 60%

27. 57.1%

29. 22.9%

31. 266.7%

33. $0.20 \times 65 = 13$

35. $0.0052 \times 1020 = 5.304$

37. $0.72x = 18$
    $x = 0.25 = 25\%$

39. $48x = 3.6$
    $x = 0.075 = 7.5\%$

41. $0.5x = 25$
    $x = 50$

43. $0.0175x = 7$
    $x = 400$

45. $300 \bullet 0.15 = 45$
    $300 - 45 = \$255$

47. $85000x = 27200$
    $x = 0.32 = 32\%$

49. $16.80(0.05) = 0.84 + 16.80 = \$17.64$

51. $1995 \bullet 0.15 = 299.25 ; 1995 - 299.25 = 1695.75$

53. $640x = 224$
    $x = 0.35 = 35\%$

55. $2890x = 9.1$
    $x = 0.003148 = 0.3\%$

57. $60x = 58.2$
    $x = 0.97 = 97\%$

59. $0.06x = 34.80$
    $x = \$580$

61. $\dfrac{4.95}{110.6} = 4.48\% ; 0.04476$